POLLUTION SCIENCE, TECHNOLOGY AND ABATEMENT

NUTRIENT POLLUTION FROM AGRICULTURAL PRODUCTION

OVERVIEW, MANAGEMENT AND A STUDY OF CHESAPEAKE BAY

POLLUTION SCIENCE, TECHNOLOGY AND ABATEMENT

Additional books in this series can be found on Nova's website under the Series tab.

Additional e-books in this series can be found on Nova's website under the eBooks tab.

POLLUTION SCIENCE, TECHNOLOGY AND ABATEMENT

NUTRIENT POLLUTION FROM AGRICULTURAL PRODUCTION

OVERVIEW, MANAGEMENT AND A STUDY OF CHESAPEAKE BAY

ARLENE WALTERS
EDITOR

nova publishers
New York

Library of Congress Cataloging-in-Publication Data

ISBN: 978-1-63485-187-9

Published by Nova Science Publishers, Inc. † New York

CONTENTS

PREFACE

Nutrients are elements essential to plant and animal growth. In agricultural production, the focus generally rests on the three primary macronutrients—nitrogen (N), phosphorus (P), and potassium (K)—because of their relative abundance in plants. As crops grow and are harvested, they gradually remove the existing nutrients from the soil and over time will require additional nutrients to maintain or increase crop yield. When nutrients are added in excess of the plants' ability to utilize them, there is an increased risk that the nutrients will enter the surrounding environment (water or air) and create environmental problems. The nutrients of primary environmental concern in agriculture are nitrogen and phosphorus. This book discusses the types and sources of nutrient pollution from agricultural production; possible environmental effects of nutrient pollution; examples of current control measures; the federal response to excess nutrients, including regulatory and incentive-based programs; and future considerations for nutrient management policy at the federal level.

In: Nutrient Pollution …
Editor: Arlene Walters

ISBN: 978-1-63485-187-9

Chapter 1

NUTRIENTS IN AGRICULTURAL PRODUCTION: A WATER QUALITY OVERVIEW*

Megan Stubbs

SUMMARY

Nutrients are elements essential to plant and animal growth. In agricultural production, the focus generally rests on the three primary macronutrients—nitrogen (N), phosphorus (P), and potassium (K)—because of their relative abundance in plants. As crops grow and are harvested, they gradually remove the existing nutrients from the soil and over time will require additional nutrients to maintain or increase crop yield. When nutrients are added in excess of the plants' ability to utilize them, there is an increased risk that the nutrients will enter the surrounding environment (water or air) and create environmental problems. The nutrients of primary environmental concern in agriculture are nitrogen and phosphorus.

One of the better-known environmental responses to high levels of nutrients is eutrophication— the enrichment of water bodies, which can promote the growth of algae. Under certain conditions, algal blooms can occur that can deplete the oxygen content of water, block sunlight to

* This is an edited, reformatted and augmented version of a Congressional Research Service publication R43919, prepared for Members and Committees of Congress, dated February 20, 2015.

other organisms, and potentially produce toxins. These harmful algal blooms can contaminate surface and drinking water supplies, potentially harming animal and human health.

Over time, through research and technological advancements, an understanding of how crops utilize nutrients and how nutrients move in the environment have led to the development of a number of best management practices (BMPs) for nutrient management. Primarily, nutrient BMPs focus on preventing or reducing the ways in which excess nutrients can enter the environment. Crop production BMPs for nutrient management generally focus on applying the right amount of nutrients, from the right source, in the right place, at the right time. BMPs for livestock operations are typically prescribed for concentrated animal feeding operations (CAFOs), where animals are raised or bred in close quarters, thus creating a concentrated source of nutrients.

Currently, few federal regulations govern the environmental impacts from agriculture. Some environmental laws specifically exempt agriculture from regulatory requirements, and others are structured so that agriculture is not addressed by most, if not all, of the regulatory impact. The primary regulatory authority protecting water resources is the Clean Water Act (CWA). Regulatory requirements for agricultural nutrients under the CWA are limited to permitting requirements for large CAFOs and the establishment of total maximum daily loads, which are pollution limits for state-identified impaired waters.

The major federal response to nutrient pollution from agriculture continues to be through research, education, outreach, and voluntary technical and financial incentives. A number of U.S. Department of Agriculture agencies provide support through education, outreach, and research, while federal funds are provided through conservation programs to help agricultural producers adopt BMPs for nutrient reduction.

As the 114th Congress reviews nutrient pollution in U.S. waterways, among the issues being discussed is how to address nutrients from agricultural sources. Whether the current balance between regulatory action and voluntary response is enough to meet environmental goals, who should bear the cost of preventing and correcting nutrient loading, and whether the tools for correction are adequate are among the issues being discussed. How these issues are resolved will have important implications for agriculture, which has taken a keen interest in water quality policy and legislation.

INTRODUCTION

United States agriculture has been touted as a model for production and a leader of innovation. The nation's intense agricultural production, however, can lead to adverse impacts to the surrounding environment. In some cases, it is the excess of basic nutrients required for plant and animal growth that can cause this degradation. Ongoing research aims to improve understanding of nutrients used and released in agricultural production, including how they interact in the environment, the damages they can cause, and ways to prevent or correct the damage.

Federal policies concerning agricultural nutrients have changed over time. Currently, few federal regulations govern the environmental impacts of nutrients from agriculture. Some environmental laws specifically exempt agriculture from regulatory requirements, and others are structured so that agriculture is not addressed by most, if not all, of the regulations. The major federal response continues to be through research, education, outreach, and voluntary technical and financial incentives to producers.

Recent events involving degraded water quality have raised questions as to whether the current federal response to agricultural nutrients is adequate or should be altered. This has prompted both administrative and congressional action.[1] This report discusses the types and sources of nutrient pollution from agricultural production; possible environmental effects of nutrient pollution; examples of current control measures; the federal response to excess nutrients, including regulatory and incentive-based programs; and future considerations for nutrient management policy at the federal level.[2]

Caveats

Agriculture is one of a number of industries that produce, use, or release nutrients that may adversely affect the environment. The lack of discussion in this report of other industries that might release excess nutrients does not imply that one industry is more or less to blame for environmental harms. The science and methods of pinpointing the exact source of excess nutrients causing environmental harm are still evolving. Regardless, whether released by agriculture, lawn care companies, or sewage treatment plants, the environmental harms of excess nutrients are much the same.

A large body of research, including numerous technical publications, explores the relationship between agriculture and nutrients. This report gives an overview of current knowledge, while further complexities of individual issues are discussed in many of the sources cited throughout the report.

Similarly, the examples used to describe best management practices and federal response should not be considered exhaustive lists. In many cases, the success or failure of a particular nutrient management practice will vary greatly by location. This report highlights examples of these practices but is not meant to imply or suggest that they can be universally applied. In addition to the federal response, the private sector, nonprofit groups, and state and local governments play active roles in the management of nutrients from agriculture, either through regulation or technical support. These nonfederal efforts are outside the scope of this report.

NUTRIENTS: A PRIMER

Nutrients are elements essential to plant growth.[3] Plant roots absorb nutrients—including water, oxygen, and others—from the soil.[4] As crops grow and are harvested, they gradually remove the existing nutrients from the soil. Over time, most soils will require additional nutrients to maintain or increase crop yield. When nutrients are added in excess of the plants' ability to utilize them, there is an increased risk that the nutrients will enter the surrounding environment (water or air) and create problems such as algal blooms (discussed below).

Plants utilize nutrients in different ways, and each plant has a different set of nutrient requirements. How, where, and when plants utilize nutrients can greatly affect the overall yield and plant production. For a farmer seeking to maximize crop yields and lower input costs, it can be critical to understand a crop's nutrient requirements.

Basic nutrients—carbon, hydrogen, and oxygen—are the most abundant elements in plants. In addition, plants utilize other nutrients commonly referred to as macronutrients and micronutrients (see *Figure 1*). In agricultural production, the focus generally rests on the three primary macronutrients—nitrogen (N), phosphorus (P), and potassium (K)—because of their relative abundance in plants. Micronutrients, while not commonly discussed, may have just as much of an effect on plant growth as macronutrients when levels are too high (toxic) or too low (deficient). This report focuses on two of the three primary macronutrients (nitrogen and phosphorus), because of the volume used in agricultural production and the relative potential for environmental harm if they are overused.[5] This is not meant to indicate that other nutrients do not also pose an environmental harm if overapplied.

Plants use nutrients in an ionic form rather than as raw elements (see *Figure 2*). Nutrients are taken up by plants in three forms:

- interception—by direct contact with the nutrient;
- mass flow—when nutrients move with water as the plant transpires;[6] and
- diffusion—when nutrients move from high to low concentration.

Interception accounts for a very low percentage of nutrients taken up by plants. Mass flow, on the other hand, is the most substantial method of nutrient movement toward a plant's roots. This is particularly important for more "mobile" nutrients (e.g., nitrogen) and less important for relatively "immobile" nutrients (e.g., phosphorus). Diffusion is most important for nutrients that are relatively immobile, have low solution concentrations, and are needed in large amounts (e.g., phosphorus and potassium).

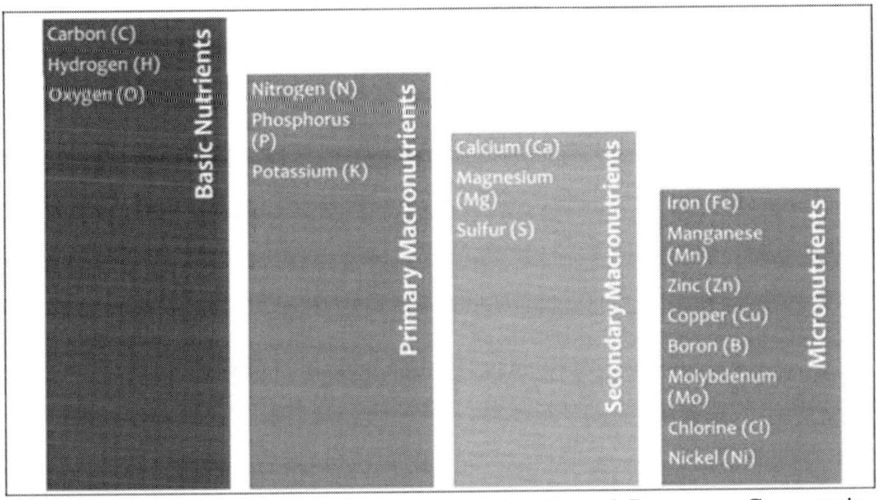

Source: U.S. Department of Agriculture (USDA), Natural Resources Conservation Service (NRCS)—4 Nutrient Stewardship, Overview of Soil Fertility, Plant Nutrition, and Nutrition Management, Education and Training Module, http://www.nutrientstewardship.com/nrcs/overview/multiscreen.html.

Notes: This figure does not represent a definitive list. Some research publications include more or fewer elements or use different categories than those presented here. Other elements, such as sodium (Na), cobalt (Co), vanadium (V), and silicon (Si), are known to be beneficial in some plants, but generally are not considered essential to most plants and therefore not included.

Figure 1. Essential Elements for Plant Growth.

Plants use nutrients at different times and at different rates throughout the growing cycle. For example, for corn, phosphorus is important early in the growing season, but not necessarily later in the season. Nitrogen is important in both the beginning and later in the growing season, before the plant begins the stem-extension phase. Various nutrients are also applied differently by farmers, based on knowledge about nutrient mobility in soils. For example, nitrogen can be applied using a broadcast method (discussed further under "Best Management Practices") because of its mobility in the soil. Phosphorus, on the other hand, is most effective when placed below the soil surface near the root system because it is generally immobile.

Adding Nutrients—Fertilizers

Fertilizers are either organic or inorganic materials applied to the soil to promote plant growth. They may either contain a specific nutrient or may be used to increase the availability of other plant nutrients. Organic fertilizers are those derived from living matter (e.g., manure) whereas inorganic fertilizers are synthetic or mined (e.g., urea as a nitrogen source or potash as a potassium source).[7]

Organic Fertilizers

The primary organic fertilizer is animal manure. Manure can be a good source of plant-available nutrients (including but not limited to nitrogen and phosphorus), as well as providing increased soil organic matter, increased water and nutrient retention in soil, and decreased soil density. The nutrient content of manure can vary greatly depending on animal type, diet, bedding, moisture content, and storage method.[8]

Prior to the development of inorganic fertilizers, organic "wastes" were a major source of nutrients for crop production. According to the U.S. Department of Agriculture (USDA), 15.8 million acres, or about 5% of all U.S. cropland, use manure as fertilizer (*Table 1*).[9] A number of factors prevent a wider use of manure fertilization. Primarily, the cost of transporting manure can be prohibitive.[10] The nutrient content of the manure, the type of crop grown and nutrients required (i.e., the nutrient needs of plants being grown may not match nutrients in readily available manure), compaction from manure application equipment, and the relative cost and availability of inorganic fertilizers, among other factors, also affect the use of organic wastes as fertilizers.

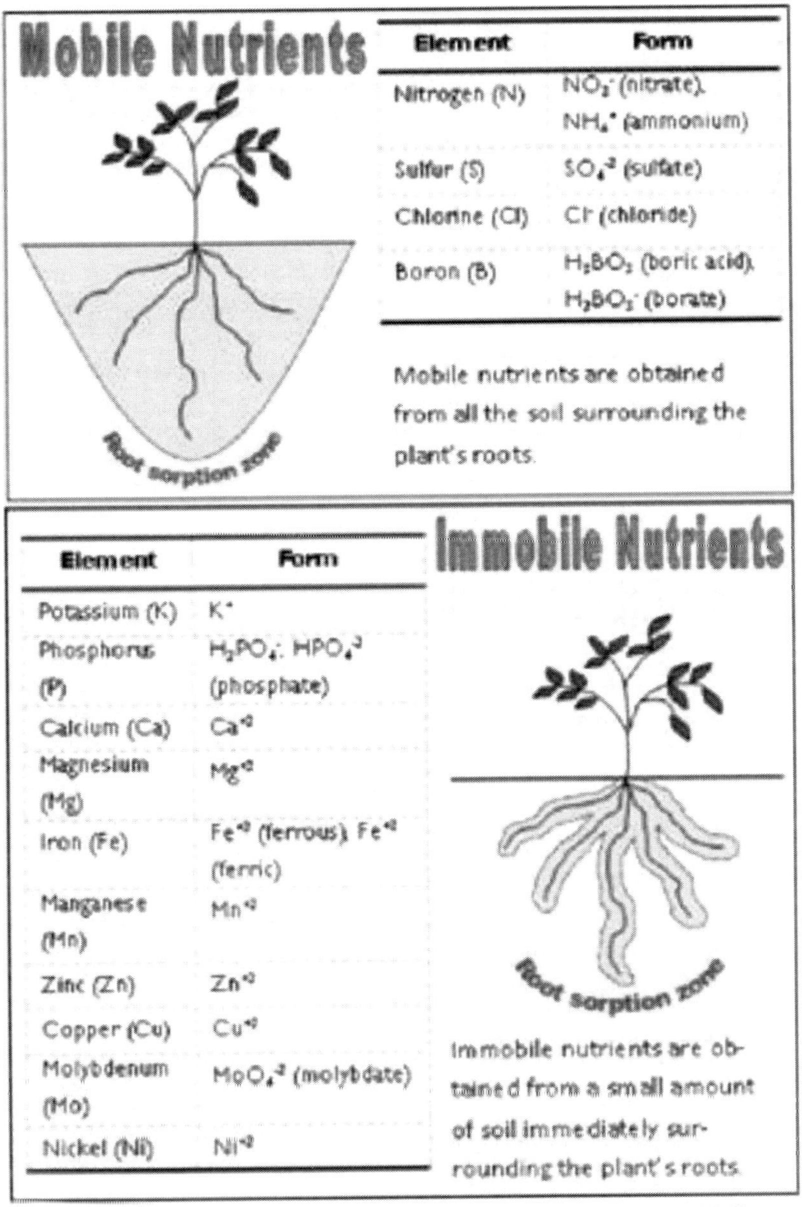

Source: CRS, modified from W. R. Raun, G. V. Johnson, and R. W. Mullen, et al., "Bray Nutrient Mobility Concept," in Soil-Plant Nutrient Cycling and Environmental Quality (Stillwater, OK: Oklahoma State University), http://soil5813.okstate.edu/BOOK.htm.

Figure 2. Relationship Between Nutrient Mobility and Plant Extraction.

Table 1. Acres Receiving Manure from Various Animal Types

(by crop and species, 2006)
Acres applied (thousand), by source of manure

Crop	Dairy	Beef Cattle	Swine	Poultry	Other[a]	All
Barley	54	36	4	4	2	100
Corn	5,612	1,617	1,161	472	224	9,086
Cotton	67	101	0	228	1	397
Oats	218	139	8	3	7	375
Peanuts	0	8	0	44	0	52
Sorghum	1	37	7	1	0	46
Soybeans	354	327	139	132	30	982
Wheat	107	250	26	12	6	401
All	6,413	2,515	1,345	896	270	11,439[b]

Source: James M. MacDonald, Marc O. Ribaudo, and Michael J. Livingston, et al., Manure Use for Fertilizer and for Energy: Report to Congress, USDA, ERS, AP037, Washington, DC, June 2009, http://www.ers.usda.gov/media/ 156155/ ap037_1_.pdf.

[a] Other includes equine, sheep, and biosolids.

[b] Other crops, such as hay and grasses, account for approximately 4.3 million acres receiving manure. These are not reflected in this table, but are in the earlier stated total of 15.8 million acres receiving manure in the United States.

From a livestock perspective, animal waste is a by-product of production that must be managed to avoid environmental harms and for the health of the animals themselves if raised in captivity. For a confined production system, manure must be utilized—typically by being collected, stored, and distributed elsewhere, likely on available crop- or pastureland. In the case of grazing livestock, the animals deposit manure directly onto pastureland, thus fertilizing the associated grassland. High stocking rates, however, may lead to an imbalance, which in turn could create an excess supply of nutrients. If not properly managed, manure can adversely impact water quality through surface runoff and erosion, direct discharges to surface waters, spills and other dry-weather discharges, and leaching into soil and groundwater. It can also result in emission to the air of particles and gases such as ammonia, hydrogen sulfide, and volatile organic chemicals.

Soil Organic Matter

Soil organic matter (SOM) is created by the cycling of organic compounds in plant, animals, and microorganisms into the soil. Essentially, SOM is soil composed of anything that once lived, including plant and animal remains, cell and tissues of soil organisms, and plant roots and soil microbes. Well-decomposed matter can form "humus," a dark brown, porous, spongy material that is essential for maintaining optimum soil physical conditions important for plant growth, water-holding capacity, and nutrient availability.11 In most soils, SOM accounts for less than 5% of the total volume but can considerably impact the nutrients available for plants.

The amount of SOM is controlled by a balance between plant and animal materials and losses from decomposition. There are a number of benefits to increasing SOM, including reduced erosion and the storing of nutrients (nitrogen, phosphorus, and sulfur). These are primarily achieved through the use of cover crops and reduced or no-till tillage systems (see the "Best Management Practices" section).

Theoretically, of the total manure produced, more could be utilized on additional cropland than what is currently utilized in order to potentially reduce the risk of environmental harms from excessive concentrations. Regions with a higher capacity for manure are largely due to the type of soils and crops grown in the area. For example, the soils (mollisols) and crops (corn) grown in the Midwest are able to utilize additional nitrogen from manure (*Figure 3*), whereas the soils (ultisols) and crops (cotton and rice) in the Southeast are able to handle more phosphorus from manure (*Figure 4*). In some cases the ability to utilize additional manure application corresponds to areas with large livestock populations. In other cases, livestock production occurs in areas with low or modest capacity to assimilate additional manure, thereby increasing the risk of excessive concentrations. When the amount of manure produced by an operation is more than the assimilative capacity of the land,[12] an excess of manure is created (*Figure 5* and *Figure 6*).[13] This can create a greater risk for potential runoff and leaching of manure nutrients and subsequent water quality issues. Geographically, areas with excess farm-level nutrients correspond to areas with increasing numbers of confined animals.[14]

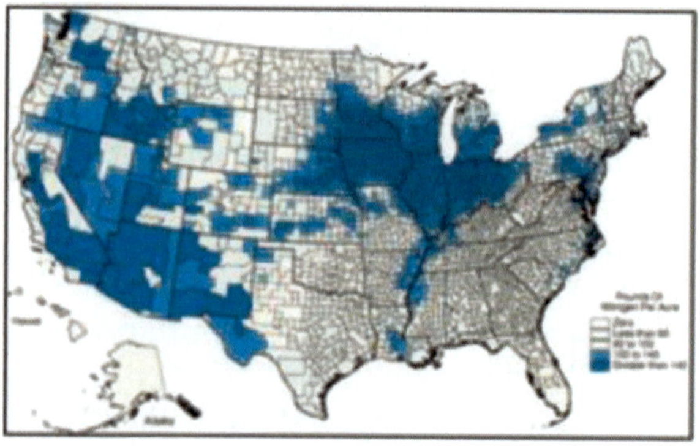

Figure 3. Capacity of Cropland to Assimilate Nitrogen.

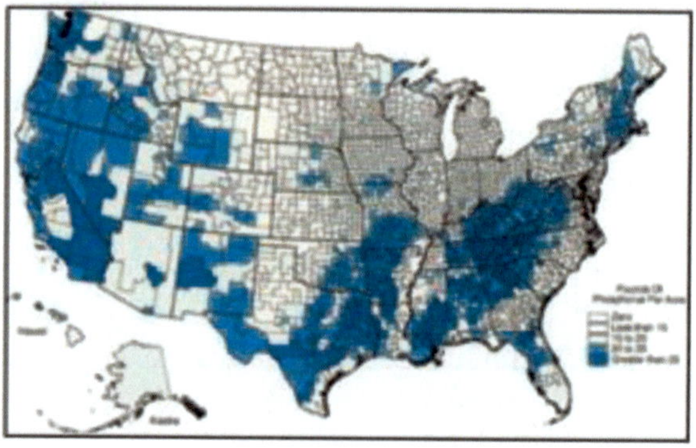

Figure 4. Capacity of Cropland to Assimilate Phosphorus.

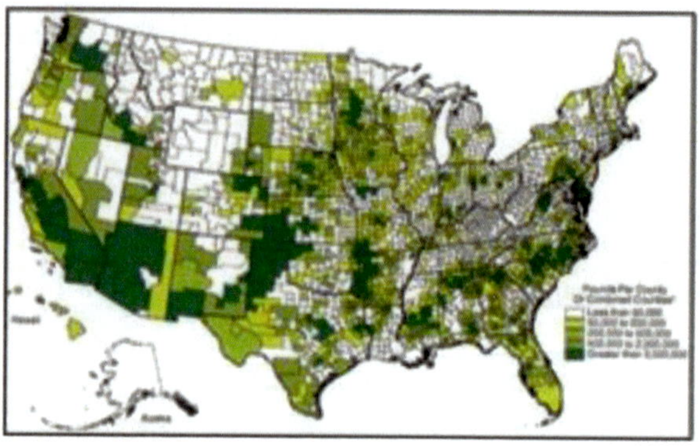

Figure 5. Manure Nitrogen in Excess of Need.

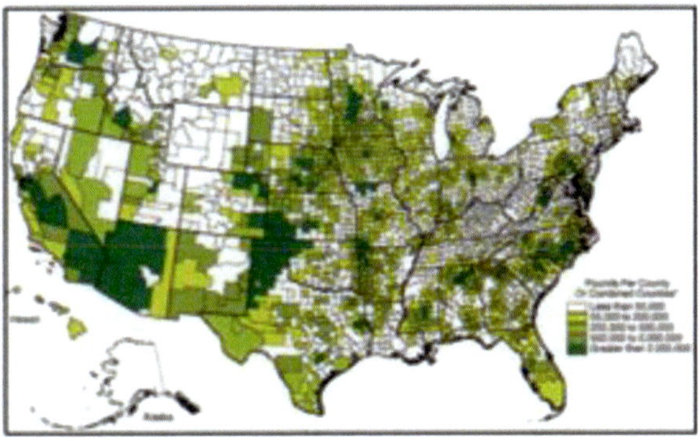

Figure 6. Manure Phosphorus in Excess of Need.

Source: Robert L. Kellogg, Charles H. Lander, and David C. Moffitt, et al., Manure Nutrients Relative to the Capacity of Cropland and Pastureland to Assimilate Nutrients: Spatial and Temporal Trends for the United States, USDA, NRCS, ERS, nps00-0579, Washington, DC, December, 2000, http://www.nrcs.usda.gov/Internet/ FSE_DOCUMENTS/nrcs143_012133.pdf.

Notes: Figures 5 and 6 assume no export from the farm.

Inorganic Fertilizers

Inorganic fertilizers consist of nutrients that are mined or created synthetically. In most cases, compared to organic fertilizers, inorganic fertilizers are more concentrated, their nutrient content is easily identifiable, and in some cases, they are more cost effective to use. This does not, however, make inorganic fertilizers any less damaging when found in excess in the environment. Similar to organic fertilizers, when applied in excess, inorganic fertilizers can be lost to the environment through volatilization into the air, leaching into groundwater, emission from soil to air, and runoff into surface water.

The use of inorganic fertilizers has changed over time and continues to outpace the use of organic fertilizers in the United States. The introduction of seed varieties that respond more favorably to specific nutrients, the use of more precise application technology, and the overall price of commercial fertilizers have driven much of this change.[15] The three primary inorganic fertilizers produced commercially are nitrogen-, phosphorus-, and potassium-based products. In general, commercial nitrogen fertilizer use has increased more than phosphorus and potassium (*Figure 7*). Corn, which uses intensive fertilizer applications, accounts for almost 40% of the total U.S. commercial fertilizer consumption, principally due to the high number of planted acres and crop requirements.[16]

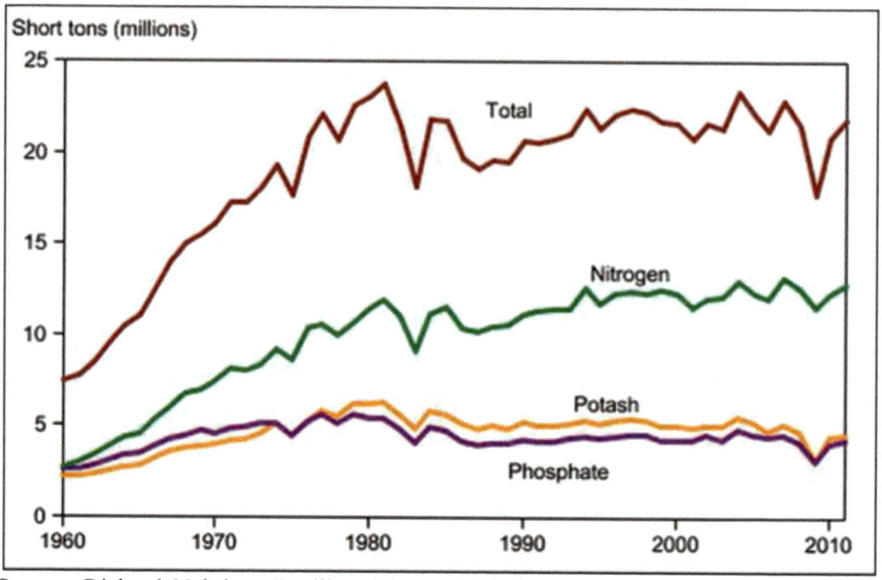

Source: Richard Nehring, Fertilizer Use and Markets, USDA, ERS, July 12, 2013, http://www.ers.usda.gov/topics/farm-practices-management/chemical-inputs/fertilizer-use-markets.aspx.

Figure 7. Fertilizer Use in U.S.Agriculture. (1960-2011).

Nitrogen

Nitrogen (N) is an abundant element, with gaseous nitrogen (N_2) accounting for 78% of the earth's atmosphere. Despite this abundance, N_2 cannot be used as a nutrient by living organisms unless converted to a useable form. This conversion occurs synthetically through the Haber-Bosch process, developed in the early 20th century, which converts "unreactive" N_2 to a more usable "reactive" form.[17] The process uses heat and pressure to combine N_2 with hydrogen from natural gas. The result is NH_3 (anhydrous ammonia), which can be used as a fertilizer directly or reacted with other compounds to form other products (*Figure 8).*[18] Approximately 74% of the NH_3 produced worldwide is for nitrogen fertilizer. Domestically, NH_3, urea (NH_3 with carbon dioxide, $CO(NH_2)_2$), and dissolved N (with water) account for 90% of total N fertilizer use.[19]

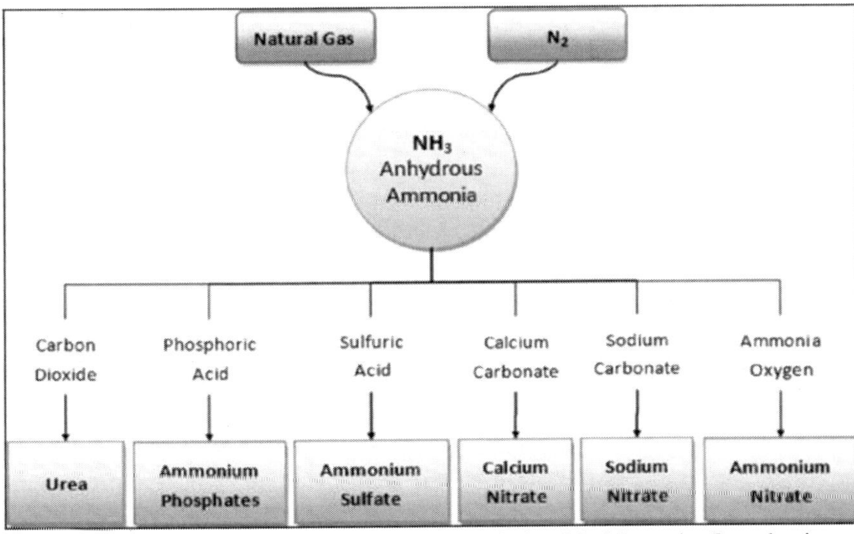

Source: Adapted from Craig C. Sheaffer and Kristine M. Moncada, Introduction to Agronomy: Food, Crops, and Environment, 2nd ed. (Clifton Park, NY: Delmar, 2012).

Figure 8. Nitrogen-Containing Fertilizers.

Phosphorus

Phosphate rock is the primary raw material used to manufacture soluble phosphorus (P) fertilizers. The sedimentary deposit is found in a number of countries, with the largest deposits in northern Africa, China, the Middle East, and the United States. Annual world phosphate rock production is projected to increase from 228 million tons in 2013 to 260 million tons in 2017.[20] In the United States, mines in Florida and North Carolina account for over 85% of domestic output. Other domestic mining operations are located in Idaho and Utah. In 2013, 95% of U.S. mined phosphate rock was used to manufacture phosphate fertilizers and animal feed supplements.[21]

Potassium

Similar to phosphorus, potassium fertilizer is primarily a mined material. Mined and manufactured salts containing soluble potassium (K) are referred to as potash.[22] The largest high-grade potash deposit in the world is in the Saskatchewan province of Canada. Domestic production occurs in Michigan, New Mexico, and Utah. In 2013, approximately 85% of potash sales in the United States were by the fertilizer industry.

EFFECTS OF NUTRIENT EXCESS ON WATER QUALITY

As nutrients cycle through the soil-plant-atmosphere continuum, some are recovered by plant uptake, some incorporated into the soil organic matter (SOM), and some precipitated as solid minerals.[23] The remainder can be transported to surface water, groundwater, and the atmosphere. The increased use of nutrients in agricultural production has increased the potential for nutrient excess and associated environmental and health impairments. Recent water and air quality concerns have brought attention to excessive nutrients in the environment and the damage they can create.[24] The nutrients of primary environmental concern in agriculture are nitrogen and phosphorus.

Water quality concerns are present across the United States for a number of reasons, including pollution from excess nutrients, heavy metals, and toxic substances, to name a few.[25] Overall, data reported by the U.S. Environmental Protection Agency (EPA) and states indicate that 44% of river and stream miles assessed by states and 64% of assessed lake acres do not meet applicable water quality standards and are impaired for one or more desired uses.[26] In 2006, EPA issued an assessment of streams and small rivers and reported that 67% of U.S. stream miles are in poor or fair condition and that nutrients and streambed sediments have the largest adverse impact on the aquatic species in these waters.[27] Agricultural production can contribute both nutrient and sediment loading[28] to waterways if not properly managed.

Nitrogen

As stated previously, nitrogen is a mobile nutrient that can occur in a variety of forms. Nitrogen is affected by chemical and biological processes that can change its form and transfer it to or from water, soil, biological organisms, and the atmosphere. The increasing use of reactive nitrogen in agriculture also increases the potential for nitrogen to be lost to the environment as ammonia (NH_3), ammonium (NH_4), nitrate (NO_3), nitrogen oxides (NO_x), and nitrous oxide (N_2O).[29]

Excess nitrogen can be transferred to water sources in a number of ways, including:

- soil erosion—either by wind or water, erosion can move soil particles containing nitrogen into surrounding waterways;

- runoff—dissolved nitrogen, either as inorganic (e.g., nitrate) or organic (e.g., manure) fertilizer, applied directly to the soil surface can "run off" the field with moving water (e.g., rain, snowmelt, or irrigation) if not incorporated into the soil; and
- leaching—water moving through the soil profile can transport dissolved nitrogen to underground water sources or through tile drains[30] to surface water.

Phosphorus

Phosphorus is an immobile nutrient and therefore is generally transferred to water through sediment-based runoff or erosion. More than 80% of phosphorus transported from cultivated lands is associated with soil particle and organic material erosion during flow events (e.g., rain or irrigation).[31] To a lesser extent, phosphorus leaching can occur through subsurface flow, primarily transported in drainage waters (e.g., through tile drains). While some soils can absorb applied phosphorus, they are not infinite sinks. Once the capacity of the soils to absorb phosphorus is exceeded, the excess will dissolve and move more freely with water.[32] Continued application of phosphorus beyond plant requirements can be a major cause of soil phosphorus saturation; both surface runoff and subsurface flow are linked to soil phosphorus concentration.

Environmental Effects

One of the better-known environmental responses to high levels of nutrients is eutrophication— the enrichment of water bodies, which can promote the growth of algae. When nutrients (e.g., nitrogen and phosphorus) and sunlight stimulate algal growth (e.g., algae, seaweed, and phytoplankton), this increases the amount of organic matter in an aquatic ecosystem over time. Algae are a natural part of the ecosystem, and most species of algae are not harmful. However, high levels of nutrients and ideal growing conditions can overfeed algae, creating algal blooms that deplete the oxygen content of water, block sunlight to other organisms, and potentially produce toxins. These harmful algal blooms (HABs) can contaminate surface and drinking water supplies, potentially harming animal and human health. Cyanobacteria (blue-green algae) and red tides are examples of HABs that produce toxins harmful to humans and animals.[33]

Algal blooms, whether toxic or not, can cause significant environmental and economic problems when the algae die. As organisms die and sink to the bottom, they are consumed (decomposed) by oxygen-dependent bacteria, depleting the water of oxygen. When this eutrophication is extensive and persistent, bottom waters may become hypoxic (depressed concentration of dissolved oxygen), or even anoxic (no dissolved oxygen). Hypoxic conditions in lakes and coastal waters can cause die-offs of fish and other aquatic life. If these conditions persist, a "dead zone" may develop in which little life exists.

Cyanobacteria

Cyanobacteria live in terrestrial, fresh, brackish, or marine water. Cyanobacterial harmful algal blooms (CyanoHABs) occur when organisms that are normally present grow exuberantly.[34] CyanoHABs are caused by a combination of conditions, including warm water temperatures, high levels of light, and abundant nutrients (primarily phosphorus and nitrogen). Nutrients play a key role, and major sources include agricultural runoff (organic and inorganic); discharges from sewage treatment plants; and storm-water runoff from lawns, streets, and elsewhere. CyanoHABs can contain various toxins that can affect the liver, skin, or nervous system. Exposure to cyanotoxins can cause a range of health effects, from mild rashes to severe illness (and rarely death) in humans. Deaths of exposed wildlife, livestock, birds, and pets have been documented worldwide. Most human exposures are thought to occur during recreational activities, such as swimming and boating; through accidental ingestion or inhalation of water; or when skin comes into contact with toxins. Exposures also can result from drinking or showering in contaminated water or eating contaminated shellfish.[35]

Why Phosphorus?

Question: If both nitrogen and phosphorus are required for a harmful algal bloom (HAB), why is phosphorus generally described as "causing" a HAB?

Answer: Nitrogen is an abundant element that is essential to the growth of aquatic life. Its abundance makes it difficult to control the atmospheric exchange of nitrogen and carbon in the water, and the fixation of atmospheric nitrogen by some blue-green algae.36 This makes phosphorus the limiting factor for a HAB. This is why control efforts typically focus on phosphorus reduction to reduce the potential for accelerated eutrophication of fresh waters.[37]

Red Tide

In the case of red tides, microscopic marine alga called *Karenia brevis* (*K. brevis*) grows quickly and creates blooms that look red or brown (hence "red tide"). *K. brevis* produces toxins called brevetoxins, which are deadly to fish and other marine organisms. Brevetoxins can become concentrated in the tissues of shellfish that feed on *K. brevis* and make those who eat these shellfish sick with neurotoxic shellfish poisoning (NSP). NSP can produce neurologic symptoms (e.g., tingling in fingers and toes) and gastrointestinal symptoms in humans. The effects of environmental exposure to brevetoxins are less well known. However, evidence suggests that air and skin exposure near red tides can result in irritation of the eyes, nose, and throat, as well as coughing, wheezing, and shortness of breath.[38]

Ciguatera

Ciguatera is another illness caused by eating fish that contain toxins produced by HABs. Ciguatera is caused by *Gambierdiscus toxicus*, a marine microalga. The toxin accumulates and can move up the food chain, for example, through large carnivorous reef fish (barracuda, black grouper, blackfin snapper, amberjack, and yellowfin grouper). Symptoms are similar to NSP and can go away in days or last for years.[39]

Human Health Effects

In addition to HABs, excess nutrients in water have other known health effects, depending on the type and condition of exposure. Conjectural evidence suggests that an excess of both nitrogen and phosphorus may contribute to infections and noninfectious pathogens, potentially causing epidemic conditions.[40] This, however, is a difficult relationship to make given the interrelated nature of humans and wildlife disease emergence.[41]

The primary route of human exposure to nitrogen is through ingestion of contaminated drinking water. The most well-known effect is methemoglobinemia, a blood disorder in which an abnormal amount of methemoglobin (a form of hemoglobin) interferes with the body's ability to release oxygen to body tissue.[42] Infants are especially susceptible to this condition, which is why it is sometimes referred to as "blue baby syndrome."[43] Other known health effects include various cancers, adverse reproductive outcomes (neural tube defects), diabetes, and thyroid conditions.[44] No comprehensive research has examined the health effects of nitrate ingestion, or whether the current regulatory limits on nitrogen in drinking water are adequately protective.[45]

Because of phosphorus's role in eutrophication and associated HABs, the environmental and health concerns are much the same as those for nitrogen. The direct health effects of excess phosphorus fertilizers are less well known than the effects of nitrogen fertilizers.

BEST MANAGEMENT PRACTICES

Nutrient management is a practice whereby nutrient cycles are kept in balance with the surrounding ecosystem. As most crop production is not considered to be a naturally occurring part of an ecosystem, how production is managed is increasingly important for the overall health of the surrounding environment and sustainability of production. Over time, through research and technological advancements, nutrient management has become increasingly sophisticated. An advanced understanding of how crops utilize nutrients and how nutrients move in the environment has led to the development of a number of best management practices (BMPs) for nutrient use. Primarily, nutrient BMPs focus on preventing or reducing the ways in which excess nutrients can enter the environment—erosion, runoff, leaching, volatilization, denitrification, and nitrification.

In many cases, the success of a BMP depends on how and where it is applied. Not all BMPs will work in every location, and more than one BMP may be required to correct a nutrient imbalance. For example, a BMP may help reduce one nutrient but do little to reduce another. Additionally, not all BMPs are fail-safe, and they can require a significant amount of time and investment to achieve a successful balance.

This section discusses select examples of nutrient BMPs that can lessen the potential harm to water quality. These examples should not be considered a comprehensive list. USDA's Natural Resources Conservation Service (NRCS) maintains national practice standards for many of these BMPs (*Table 2*); however, state and local regulations may affect how the standards apply locally.

Table 2. Select NRCS Practice Standards Related to Nutrient Reduction and Water Quality

Name	Practice Code[a]	Link
Animal mortality facility	316	http://www.nrcs.usda.gov/Internet/FSE_DOCUMENTS/nrcs143_026388.pdf
Agrichemical handling facility	309	http://www.nrcs.usda.gov/Internet/FSE_DOCUMENTS/stelprdb1263506.pdf
Composting facility	317	http://www.nrcs.usda.gov/Internet/FSE_DOCUMENTS/nrcs143_026122.pdf
Conservation crop rotation	328	http://www.nrcs.usda.gov/Internet/FSE_DOCUMENTS/stelprdb1263170.pdf
Constructed wetland	656	http://www.nrcs.usda.gov/Internet/FSE_DOCUMENTS/nrcs143_025770.pdf
Contour buffer strips	332	http://www.nrcs.usda.gov/wps/PA_NRCSConsumption/download?cid= nrcs143_026249&ext=pdf
Contour farming	330	http://www.nrcs.usda.gov/Internet/FSE_DOCUMENTS/stelprdb1241315.pdf
Cover crop	340	http://www.nrcs.usda.gov/Internet/FSE_DOCUMENTS/stelprdb1263176.pdf
Drainage water management	554	http://www.nrcs.usda.gov/Internet/FSE_DOCUMENTS/nrcs143_026409.pdf
Feed management	592	http://www.nrcs.usda.gov/Internet/FSE_DOCUMENTS/stelprdb1046856.pdf
Filter strip	393	http://www.nrcs.usda.gov/Internet/FSE_DOCUMENTS/stelprdb1241319.pdf
Grassed waterway	412	http://www.nrcs.usda.gov/Internet/FSE_DOCUMENTS/stelprdb1263177.pdf
Irrigation water management	449	http://www.nrcs.usda.gov/Internet/FSE_DOCUMENTS/stelprdb1263179.pdf
Nutrient management	590	http://www.nrcs.usda.gov/Internet/FSE_DOCUMENTS/stelprdb1046433.pdf
Residue and tillage management, no till	329	http://www.nrcs.usda.gov/Internet/FSE_DOCUMENTS/stelprdb1249901.pdf
Residue and tillage management, reduced till	345	http://www.nrcs.usda.gov/Internet/FSE_DOCUMENTS/stelprdb1251402.pdf
Short-term storage of animal waste and byproduct	318	http://www.nrcs.usda.gov/Internet/FSE_DOCUMENTS/stelprdb1263507.pdf
Stream crossing	578	http://www.nrcs.usda.gov/Internet/FSE_DOCUMENTS/stelprdb1046932.pdf

Table 2. (Continued)

Name	Practice Code[a]	Link
Stripcropping	585	http://www.nrcs.usda.gov/Internet/FSE_DOCUMENTS/nrcs143_026280.pdf
Terrace	600	http://www.nrcs.usda.gov/Internet/FSE_DOCUMENTS/stelprdb1263187.pdf
Waste storage facility	313	http://www.nrcs.usda.gov/Internet/FSE_DOCUMENTS/nrcs143_026465.pdf
Waste treatment	629	http://www.nrcs.usda.gov/Internet/FSE_DOCUMENTS/stelprdb1144479.pdf
Waste treatment lagoon	359	http://www.nrcs.usda.gov/Internet/FSE_DOCUMENTS/nrcs143_026002.pdf

Source: CRS from USDA, NRCS, "National Conservation Practice Standards," http://www.nrcs.usda.gov/wps/ portal/nrcs/main/national/technical/cp/ncps/.

Notes: This is not intended to be a comprehensive list. In some cases, additional practices are required to complete a conservation system. For example, a manure management system may consist of a number of practices (e.g., waste storage facility [313] and nutrient management [590]), including some not on this list (e.g., waste transfers [634], pumping plants [533], irrigation water conveyance [428], and irrigation system sprinkler [422]).

[a] Practice codes refer to an NRCS numbering system and are included for reference purposes.

Crop Production

Crop production BMPs for nutrient management generally focus on the "4Rs"— applying the *right* amount of nutrients, from the *right* source, in the *right* place, at the *right* time. Other associated BMPs try to prevent nutrients and sediment from leaving the field and entering waterways. Select examples of crop production BMPs for water quality improvement include:

Nutrient Diagnosis and Testing. Fertilizer recommendations are often based on nutrient diagnostic methods, such as soil testing (*Figure 9*), plant analysis, and canopy sensing. These activities measure the amount of nutrients present and help determine additional need. The results, combined with data on expected yields and field conditions, help producers minimize excessive nutrient release into the environment.

Source: Tim McCabe, USDA, NRCS.

Note: A soil sample is drawn early in the crop year to test for N availability in the soil and again later in the spring to determine additional need.

Figure 9. Soil Sample.

Source: CRS.

Note: An example of soybeans in corn residue using a no-till system. Also an example of a crop rotation between corn, soybeans, and wheat

Figure 10. No-Till System and Crop Rotation.

Planning. Nutrient management planning calculates the amount of nutrients required for maximum yield, while minimizing the overall environmental impacts of nutrient use. NRCS defines nutrient management as the management of the "4Rs."[46]

Tillage Systems. The use of no-till (*Figure 10)* or reduced till systems can increase soil organic matter, reduce nutrient use, reduce sediment loss, enhance ground-water retention, and reduce runoff and leaching of nutrients. No-till47 systems maintain most of the crop residue on the soil surface by not using traditional full-width tilling methods throughout the year. Reduced till, also referred to as mulch till, disturbs the majority of the soil's surface with noninversion tillage methods48 and uniformly spreads residue on the soil surface.

Cover Crops. Legume crops and cover crops can provide nitrogen through biological fixation and nutrient recycling. The use of cover crops helps to sequester nutrients, primarily nitrogen; reduce soil run-off; and improve nutrient use efficiency.

Crop Rotation. Crop rotations are planned sequences of two or more crops grown on the same ground over a period of time. Whether rotating with other commodity crops or perennials, the use of crop rotations has been known to improve nutrient cycling and reduce energy needs (*Figure 10*).

Stripcropping. The use of grass or closegrowing crops alternated with row crop production can slow runoff, reduce erosion, and increase filtration. This can reduce the loss of nitrates and soluble phosphorus (see *Figure 11*).

Application Method. How a plant uses nutrients will determine where nutrients should be placed for most efficient uptake. The application method is determined based on crop type, nutrient diagnosis and testing, and planning. Different application methods are available based on the time of application (see *Figure 12*).

Application Timing. How a plant uses nutrients will determine when nutrients should be applied for most efficient uptake. In some cases, this might mean more than one application at different times (often referred to as split application).

Inhibitors. Nitrification inhibitors maintain ammonium longer by eliminating a bacterium where the ammonium is present, thereby delaying denitrification and reducing leaching potential. Urease inhibitors allow urea to be retained in the soil longer, thereby reducing nitrogen volatilization.[49]

Application Amount. The amount of nutrient application should be based on the results of nutrient diagnosis and testing; realistic expected yields and yield history; and residual nutrients from manure, legumes, and irrigation water applied to reduce overapplication.

Source: Tim McCabe, USDA, NRCS
Note: Alternating strips of alfalfa with corn on the contour.

Figure 11. Stripcropping.

Filter and Buffer Strips. A strip of vegetation located near the edge of the field along streams, lakes, wetlands, and other adjacent waters in order to trap sediment and denitrify residual nitrates in subsurface flow.

Constructed Wetlands. Surface or subsurface drainage tile water may be channeled through artificially created wetlands to provide nutrient filtration.

Irrigation Management. Coordinating the use of irrigation water and nutrient application can impact nutrient uptake, runoff, and leaching. Other irrigation system practices, such as tailwater recovery, can also assist in capturing residual nutrient loss.

Terraces and Grassed Waterways. Terraces are a series of ridges and channels constructed across a slope to intercept sediment and nutrient runoff. Grassed waterways are channels with established vegetation that reduces runoff (see *Figure 13*).

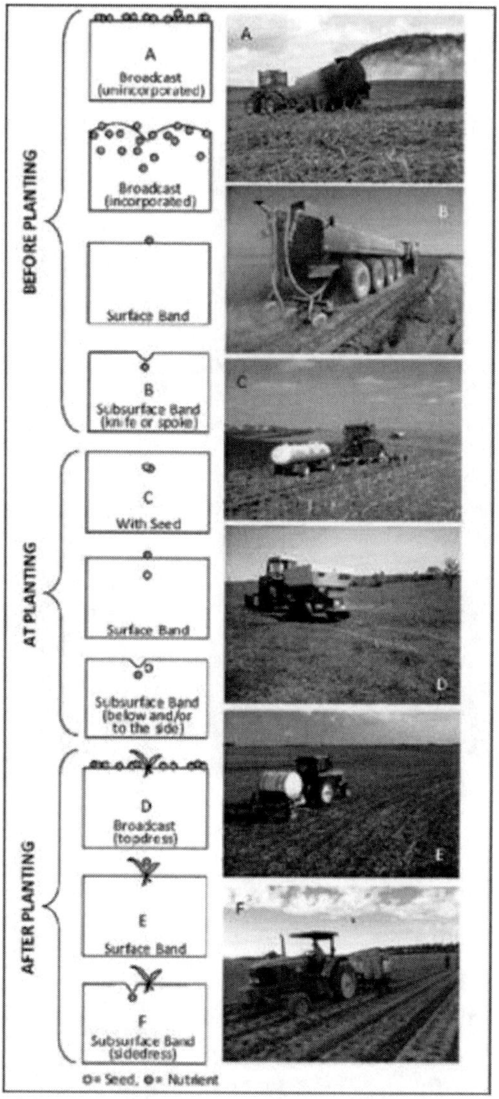

Source: CRS adapted from John L. Havlin, Samuel L. Tisdale, and Werner L. Nelson, et al., Soil Fertility and Fertilizers: An Introduction to Nutrient Management, 8th ed. (Pearson, 2014), figs. 10-15. Photos A-E from USDA, NRCS. Photo F from the University of Arizona, College of Agriculture and Life Sciences/College of Engineering, https://www.cals.arizona.edu/abe/featured/rotary-soilinjection-approaches-improved-side-dress-applications.

Note: Not all application methods represent a BMP.

Figure 12. Select Examples of Nutrient Application Methods.

Source: Jeff Vanuga, USDA, NRCS.
Note: Shows contour terraces and grassed waterways.

Figure 13. Terraces and Grassed Waterways.

Animal Agriculture

Best management practices for livestock operations are typically prescribed for concentrated animal feeding operations where animals are raised or bred in close quarters, thus creating a concentrated source of nutrients. Other BMPs for grazing operations can reduce the potential for nutrient and sediment contamination by controlling animals' movements within grazing sites.

Nutrient Testing. The nutrient content of manure can vary greatly. Nutrient testing of manure can inform feed and fertilizer application recommendations. Testing results can help minimize nutrient release into the surrounding environment.

Planning. Animal waste planning helps producers determine their current waste and nutrient production and distribution capacities. Balancing the production and distribution of waste is critical to preventing excess nutrient release.

Feed Management. Modifying animal feed diets can reduce the nutrient content of the manure.

Carcass Disposal. Improper disposal of livestock and poultry carcasses can impact surface and groundwater resources. Animal mortality facilities vary and can consist of composters, refrigeration, incinerators, or burial pits, among others.

Constructed Wetlands. Constructed wetlands can provide nutrient filtration from wastewater and contaminated runoff from livestock and agricultural processing facilities. Constructed wetlands generally use hydrophytic vegetation (i.e., grows in water) in a shallow basin where the contaminated water, both entering and exiting the wetland, is controlled (*Figure 14*).

Waste Storage and Handling. The storage of manure waste can allow for greater flexibility in nutrient application and timing for crop production. Practices such as short-term storage— consisting of plastic sheeting—or longer-term storage facilities (e.g., pits or lagoons) are generally pursued as part of a larger nutrient management plan.

Lagoon. A lagoon is a pond-like earthen basin that provides biological treatment and long-term storage of animal waste. Generally manure is diluted with water, decomposed through a biological process, and then used in a separate form such as irrigation liquid. The biological reaction is achieved by either anaerobic bacteria (inhibited by oxygen) or aerobic bacteria (requiring oxygen).

Stream Crossing. A stabilized area constructed across a stream or waterway to provide access for people or livestock while reducing sediment and nutrient loading in the stream.

Federal Response to Agricultural Nutrients

Currently, few federal regulations govern the environmental impacts from agriculture. Some environmental laws specifically exempt agriculture from regulatory requirements, and others are structured so that agriculture is not addressed by most, if not all, of the regulatory impact. The major federal response continues to be through research, education, outreach, and voluntary technical and financial incentives.

Source: Tim McCabe, USDA, NRCS.

Note: A series of hillside terraces form constructed wetlands that filter water from a hog operation.

Figure 14. Constructed Wetland.

Regulation—Clean Water Act[50]

Federal environmental law does not regulate all agricultural activities.[51] In terms of environmental impacts, the primary regulatory focus has been on protecting water resources and is governed by the Clean Water Act (CWA). As with many environmental regulations affecting industries, regulations affecting agriculture have been and continue to be controversial and draw congressional attention.

The CWA provides one exception to policies that generally exempt agricultural activities—and specifically the livestock industry—from environmental rules. The law protects water quality through a combination of ambient water quality standards established by states, limits on effluent discharges, and permits. The regulatory structure of the CWA distinguishes between "point sources" (e.g., manufacturing and other industrial facilities that are regulated by discharge permits) and "nonpoint sources" (pollution that occurs in conjunction with surface erosion of soil by water and surface runoff of rainfall or snowmelt from diffuse areas, such as farm and ranch land). Most agricultural activities are considered to be nonpoint sources, since they do not discharge wastes from pipes, outfalls, or similar conveyances. Pollution from nonpoint sources is generally governed by state water quality planning provisions of the act.

Concentrated Animal Feeding Operation (CAFO) Permits[52]

The CWA prohibits the discharge of pollutants from any point source to waters of the United States unless authorized under a permit that is issued by EPA or a qualified state, and the act expressly defines concentrated animal feeding operations (CAFOs) as point sources.[53] Permits limiting the type and quantity of pollutants that can be discharged are derived from effluent limitation guidelines promulgated by EPA under the National Pollutant Discharge Elimination System (NPDES) program. In 2003, EPA revised regulations that were first promulgated in the 1970s defining the term CAFO for purposes of permit requirements and specifying effluent limitations on pollutant discharges from regulated feedlots. The 2003 rules were challenged in federal court (*Waterkeeper Alliance et al. v. EPA*, 399 F.3d 486 (2nd Cir. 2005)), and parts of the regulations were remanded to EPA for revision and clarification. As a result, EPA issued revised regulations in 2008.[54]

The 2008 CAFO rule applies to approximately 15,300 of the largest animal feeding operations that confine cattle, dairy cows, swine, sheep, chickens, laying hens, and turkeys, or less than 10% of all animal confinement facilities in the United States. The rule details requirements for permits, annual reports, and development of plans for handling manure and wastewater. The rule contains a performance standard that prohibits discharges from regulated CAFOs except in the event of wastewater or manure overflows or runoff from an exceptional 25-year, 24-hour rainfall event. Parts of the rule are intended to control land application of animal manure and wastewater.

Total Maximum Daily Load (TMDL)[55]

Section 303(d) of the CWA requires states to identify waters that are impaired by pollution, even after application of pollution controls. For those waters, states must establish a total maximum daily load (TMDL) to ensure that water quality standards can be attained. A TMDL is essentially a pollution budget, a quantitative estimate of what it takes to achieve standards, setting the maximum amount of pollution that a waterbody can receive without violating standards. If a state fails to do this, EPA is required by the CWA to make its own TMDL determination for the state.

Throughout the United States, more than 20,000 waterways are known to be violating applicable water quality standards and require a TMDL. Lawsuits have been brought with the intention of pressuring EPA and states to develop TMDLs because the waters have been identified as being impaired (that is, not meeting applicable water quality standards). The Chesapeake Bay TMDL is the largest single TMDL developed to date.

Since 1995, EPA and states have developed a total of 67,726 TMDLs, addressing 70,901 causes of water quality impairment in U.S. waters. Of the 67,726 total TMDLs, 5,927 (8.75%) addressed nutrient impairments, and 3,904 (5.8%) addressed sediment impairments.[56] According to EPA, 42,316 waterbodies remain impaired in the United States, due to 74,594 causes of impairment (i.e., a waterbody may be impaired by more than one pollutant). TMDLs are to be established for these waterbodies. Of the total known causes of current impairment, 7,688 (10.3%) are known to be caused by nutrients, and 6,466 (8.7%) are known to be caused by sediment.[57] Generally it is not possible to determine how many of the nutrient impairments for which TMDLs are to be developed are caused by agriculture or other possible sources until the TMDL is developed by a state. The identification and listing of impaired waters are done by states.

Production Support

In large part, the federal response to nutrient pollution from agriculture is through voluntarily adopted technical and financial assistance. A number of USDA agencies provide support through education, outreach, and research, while federal funds are provided through conservation programs to adopt BMPs for nutrient reduction.

Technical Assistance, Education, and Outreach

USDA offers technical support to producers through direct assistance and research. Nutrient management planning assistance is available at no cost through NRCS. Agency resources and planning techniques draw on publicly reviewed and scientifically based principles. Various conservation practices (the BMPs described above) may be prescribed, based on needs and resource goals. Unless the nutrient management plan was created for a CAFO to meet the regulatory requirements under the CWA NPDES program, implementation of a nutrient management plan and associated BMPs are voluntary. Other state- and local-level programs (e.g., tax credits, pollution violation compliance, cost-share agreement, or local ordinance) may require implementation of a nutrient management plan, but this varies greatly by state.

Nutrient management education and outreach is primarily handled through the land-grant university system and the extension programs of the National Institute of Food and Agriculture (NIFA).[58]

Financial Assistance

In addition to technical assistance, NRCS also administers a number of conservation programs that offer financial assistance for nutrient management. According to NRCS, 8.6% of all acres receiving NRCS conservation assistance for water quality between FY2005 and FY2012 were for practices that support nutrient management.[59]

Many of the programs that fund nutrient management BMPs are authorized in omnibus farm bills, the most recent of which is the Agricultural Act of 2014 (P.L. 113-79). The Environmental Quality Incentives Program (EQIP) is the largest farm bill program that funds nutrient management practices, including conservation activity plans for nutrient management as well as financial assistance to implement other nutrient management practices recommended in the nutrient management plan (e.g., cover crops, buffer strips, waste lagoons). Other farm bill programs, such as the Conservation Stewardship Program (CSP), provide incentives encouraging producers to adopt additional nutrient-reducing practices. Land retirement and easement programs, such as the Conservation Reserve Program (CRP) and the Agricultural Conservation Easement Program (ACEP), remove land from production and establish resource-conserving vegetation for wildlife and water quality benefits.[61]

The Clean Water Act Section 319 Program authorizes grants to states, territories, and tribes to help address national water quality challenges posed by nonpoint sources of pollution, such as runoff from farmland, forests, and city streets.[62] State Section 319 programs generally address nonpoint source pollution from a variety of sectors, not just agriculture. According to EPA, states use their Section 319 funding, along with a state match (40% match is required), to implement statewide, non-regulatory programs that promote implementation on a widespread basis (e.g., promote broad use of nutrient management in agriculture).[63]

Conservation Activity Plans

Amendments in the 2008 farm bill60 allow NRCS to use financial assistance to pay for conservation plan development. The plans are referred to as Conservation Activity Plans (CAPs), of which two are for nutrient management—#102 comprehensive nutrient management plan (CNMP, for CAFOs and AFOs) and #104 nutrient management plan (non-livestock, crop production). EQIP can also pay for plan implementation, but this is handled separately from the CAPs. The main practice standard for nutrient management implementation is the 590 practice standard (see Table 2). CAP #102 or #104 describes which nutrients need to be managed and how they could be managed, whereas the 590 nutrient management practice will fund the activities required to manage them (e.g., soil testing, design, and labor).

Research and Monitoring

USDA conducts nutrient management research both through its in-house research agencies— Agricultural Research Service (ARS) and the Economic Research Service (ERS)—and through funds provided to states and localities by NIFA. The ARS conducts numerous research projects within 17 national programs, five of which support nutrient management-related research.[64] The ERS, which conducts economic research on policy issues, has produced a number of publications that analyze the societal and economic impacts of nutrient use in the United States from the standpoint of both production and environmental quality.

USDA also participates in a multi-agency effort led by NRCS, called the Conservation Effects Assessment Project (CEAP). The purpose of CEAP is to quantify the environmental effects of conservation practices and programs and develop the science base for managing the agricultural landscape for environmental quality. Project findings are used to guide USDA conservation policy and program development and potentially to assist producers with making more informed conservation decisions. To date, a number of CEAP publications focus on nutrient management.[65]

The U.S. Geological Survey (USGS), as part of its National Water Quality Assessment Program, is conducting studies on the transport and fate of agricultural nutrients in agricultural settings across the country. Early study results highlight how environmental processes and agricultural practices interact to affect the movement and transformation of agricultural chemicals in the environment. The study covers surface water, groundwater, the unsaturated zone, the streambed, and the atmosphere, as well as the pathways that interconnect these compartments. In an attempt to make the findings nationally relevant, the study areas represent major agricultural settings, such as diverse irrigated crop systems in Western states and corn and soybean row crop systems in the Midwest.[66]

POLICY QUESTIONS

Despite advancements made through research, education, and funding, environmental problems with excess nutrients from agriculture remain throughout the United States. As policymakers consider these issues, additional questions may be asked.

- *Regulatory-Voluntary Balance*: Currently, few environmental regulations govern nutrients from agriculture; instead, a more voluntary incentive-based approach is used. What benefits are achieved through the voluntary approach, and how could they be strengthened? Could additional benefits be achieved through regulation, and at what cost? What is an acceptable balance between the two?
- *Federal-State-Local Balance*: In many cases, the enforcement of federal laws that address environmental pollution is delegated to the states. Does this model work? Should more or less oversight or action occur, and at what level? How can regional and watershed planning impact the overall effectiveness of the current environmental laws?
- *Use of Current Resources*: Congress authorizes and appropriates billions of dollars for federal conservation efforts each year. Demand for these resources continues to outpace supply. How are current federal resources for conservation being utilized? How effective are

they, and how can they be improved? Are federal leveraging programs—such as the Regional Conservation Partnership Program, which leverages federal funds with private funds—an effective way to extend the federal dollar? Agencies and stakeholders have suggested "targeting" limited resources to areas of greatest need. To what extent is targeting effective, and what barriers to targeting exist?

- *The Source of the Problem*: The exact sources and causes of nutrient pollution can vary greatly by location. Case study analysis and modeling can assist with identifying sources at the local level but provide little in the way of nationwide conclusions. What additional research or monitoring efforts are required to identify sources of nutrient pollution? At what level should these efforts occur— field level, state level, nationally?
- *Effectiveness of the Solution*: Conservation BMPs are frequently held up as being the solution to agricultural nutrient pollution. What scientific evidence supports the use of these BMPs? Can additional research and innovation provide other "tools" in the response "toolbox"?
- *Value Versus Cost of Conservation*: Agricultural producers seek to maximize profit, while simultaneously minimizing cost. Does the value of conservation BMPs outweigh the cost? Are savings and efficiencies (if any) adequately communicated to producers to increase adoption?
- *Consequences of Actions*: Nutrient pollution may or may not directly impact the polluters themselves. According to ERS, the cost of removing nitrates from U.S. drinking water supplies is over $4.8 billion per year. The bulk of this cost is borne by large water utilities and then passed on to consumers. It is estimated that if the agricultural industry were required to pay based on its contribution to nitrate loading, agriculture's share would be about $1.7 billion per year.[67] Who should bear the cost of nutrient loading? Can education and outreach connect on-farm actions to potential off-farm results?

CONCLUSION

Increased demand to feed a growing population is frequently cited as the reason for additional agricultural output. Applying additional nutrients for increased yields and managing nutrients from animal production will likely continue if these demands are realized. This will only elevate the importance of managing these additional nutrients correctly or potentially repeating and exacerbating the environmental degradation that has occurred in the past.

Agriculture is one of a number of industries that produces, uses, or releases nutrients that may adversely affect the environment. The exact source of water impairments is frequently not identified until major ecological events prompt additional research. Even then, the nature of how nutrients move through the environment can make their origin difficult to track. It is agriculture's use of nutrients and large land-use presence across the United States that have brought attention to how the industry utilizes and manages nutrients. And while best management practices exist to eliminate or reduce nutrients from entering waterways, they are not required, they may not be effective or affordable in all locations, and the success will depend on how and where they are applied.

Some advocate for the additional regulation of agricultural nutrient activities, while others seek to expand incentive-based resources. The track record of success and failure for both options varies, making the likely solution a balance of the two. It is this exact balance that Congress continues debate through the oversight of current federal activities and consideration of new or modified initiatives. Research, data, and monitoring continue to impact the debate, but perhaps not as much as larger water impairments that affect public health and safety.

End Notes

[1] U.S. Congress, Senate Committee on Agriculture, Nutrition, and Forestry, Farmers and Fresh Water: Voluntary Conservation to Protect Our Land and Waters, 113th Cong., 2nd sess., December 3, 2014.

[2] While distinctly related, air quality issues are not discussed in detail but may be mentioned where applicable.

[3] Nutrients are also essential to animal health; however, they are discussed in this report in the context of nutrients as a by-product of livestock production (e.g., manure, bedding, etc.) rather than an input.

[4] Some soils are naturally fertile because they contain minerals high in key elements, while others include high levels of organic matter and thus high levels of key nutrients. There are a number of physical, chemical, and biological soil properties that affect the nutrient availability for plants.

[5] Potassium is a mineral naturally found in soils. It is essential for plant growth and is frequently found in amounts that exceed the amount used by crops in a given season. Unlike other

minerals, however, potassium does not act as a pollutant in the environment and therefore is not discussed in depth in this report.

[6] Transpiration is the process by which moisture is carried through plants from roots to small pores on the underside of leaves, where it changes to vapor and is released to the atmosphere. Transpiration is essentially evaporation of water from plant leaves. For more information, see U.S. Geological Survey, The Water Cycle, "Transpiration – The Water Cycle," http://water.usgs.gov/edu/watercycletranspiration.html.

[7] Craig C. Sheaffer and Kristine M. Moncada, Introduction to Agronomy: Food, Crops, and Environment, 2nd ed. (Clifton Park, NY: Delmar, 2012).

[8] Determining the nutrient levels of manure can be done through sampling and testing prior to application. Because the nutrient content of manure is usually the largest unknown and first step to preventing overapplication, sampling and testing is strongly encouraged as a best management practice (discussed further in "Best Management Practices" below).

[9] James M. MacDonald, Marc O. Ribaudo, and Michael J. Livingston, et al., Manure Use for Fertilizer and for Energy: Report to Congress, USDA, Economic Research Service (ERS), AP037, Washington, DC, June 2009, http://www.ers.usda.gov/media/156155/ap037_1_.pdf.

[10] One exception to this is for poultry litter, which is dry (and typically less costly to transport) and produced on farms that typically have no crop production. Thus, crops such as peanuts and cotton, which are grown in areas near poultry production (e.g., the Southeast), tend to rely on poultry litter transported from off the farm.

[11] USDA, NRCS, National Soil Survey Center, Soil Quality Indicators: Organic Matter, Soil Quality Information Sheet, Lincoln, NE, April 1996.

[12] The assimilative capacity refers to the estimated amount of nutrients taken up and removed at harvest for cropland and the amount that could generally be applied to pastureland without accumulating nutrients in the soil.

[13] Robert L. Kellogg, Charles H. Lander, and David C. Moffitt, et al., "Manure Nutrients Relative to the Capacity of Cropland and Pastureland to Assimilate Nutrients: Spatial and Temporal Trends for the United States," USDA, NRCS, ERS, nps00-0579, Washington, DC, December, 2000, http://www.nrcs.usda.gov/Internet/FSE_DOCUMENTS/ nrcs143_012133.pdf.

[14] Ibid.

[15] Richard Nehring, Fertilizer Use and Markets, USDA, ERS, July 12, 2013, http://www.ers.usda.gov/topics/farmpractices-management/chemical-inputs/fertilizer-use-markets.aspx.

[16] Ibid.

[17] Marc Ribaudo, Jorge Delgado, and LeRoy Hansen, et al., Nitrogen in Agricultural Systems: Implications for Conservation Policy, USDA, ERS, ERR-127, Washington, DC, September 2011.

[18] Craig C. Sheaffer and Kristine M. Moncada, Introduction to Agronomy: Food, Crops, and Environment, 2nd ed. (Clifton Park, NY: Delmar, 2012).

[19] John L. Havlin, Samuel L. Tisdale, and Werner L. Nelson, et al., Soil Fertility and Fertilizers: An Introduction to Nutrient Management, 8th ed. (Boston: Pearson, 2014).

[20] Stephen M. Jasinski, Phosphate Rock, USGS (U.S. Geological Survey), Mineral Commodity Summaries, February 2014, http://minerals.usgs.gov/minerals/pubs/commodity/phosphate_rock/mcs-2014-phosp.pdf.

[21] Ibid.

22 Stephen M. Jasinski, Potash, USGS (U.S. Geological Survey), Mineral Commodity Summaries, February 2014, http://minerals.usgs.gov/minerals/pubs/commodity/potash/mcs-2014-potas.pdf.

[23] John L. Havlin, Samuel L. Tisdale, and Werner L. Nelson, et al., Soil Fertility and Fertilizers: An Introduction to Nutrient Management, 8th ed. (Boston: Pearson, 2014).

[24] While distinctly related, air quality issues are not discussed in detail in this report.

[25] For additional information, see CRS Report R43867, Water Quality Issues in the 114th Congress: An Overview.

[26] U.S. Environmental Protection Agency, National Water Quality Inventory: Report to Congress, 2004 Reporting Cycle, EPA 841-R-08-001, January 2009, http://water.epa.gov/lawsregs/guidance/cwa/305b/2004report_index.cfm.

[27] U.S. Environmental Protection Agency, Wadeable Streams Assessment: A Collaborative Survey of the Nation's Streams, EPA 841-B-06-002, December 2006, http://www.epa.gov/owow/streamsurvey/.

[28] Erosion caused by water is the detachment and transport of soil particles by rainfall or irrigation. These soil particles flow through channels to associated waterways and are referred to as a sediment load. This is a naturally occurring process, but it can be accelerated with heavy water events and an unstable soil surface.

[29] Marc Ribaudo, Jorge Delgado, and LeRoy Hansen, et al., Nitrogen in Agricultural Systems: Implications for Conservation Policy, USDA, ERS, ERR-127, Washington, DC, September 2011.

[30] Tile drains are a subsurface drainage system that removes excess water from the soil profile through a series of perforated tubes.

[31] A. N. Sharpley, T. Daniel, and T. Sims, Agricultural Phosphorus and Eutrophication, Second Edition, USDA, Agricultural Research Service, ARS-149, September 2003, http://www.ars.usda.gov/is/np/Phos&Eutro2/ agphoseutro2ed.pdf.

[32] Joseph L. Domagalski and Henry Johnson, Phosphorus and Groundwater: Establishing Links Between Agricultural Use and Transport to Streams, U.S. Geological Survey, Fact Sheet 2012-3004, January 27, 2012, http://pubs.usgs.gov/ fs/2012/3004/.

[33] For more information, see CRS Report IN10131, Harmful Algal Blooms and Drinking Water.

[34] Centers for Disease Control and Prevention, Harmful Algal Blooms (HABs), http://www.cdc.gov/nceh/hsb/hab/ default.htm.

[35] U.S. Environmental Protection Agency, Cyanobacteria/Cyanotoxins, http://www2.epa.gov/nutrient-policy-data/ cyanobacteriacyanotoxins.

[36] When salinity increases, as in estuaries, nitrogen generally becomes the element controlling aquatic productivity. David K. Mueller and Dennis R. Helsel, Nutrients in the Nation's Waters—Too Much of a Good Thing? U.S. Geological Survey, Circular 1136, January 11, 2013, http://pubs.usgs.gov/circ/circ1136/.

[37] Studies indicate that increased nitrogen and phosphorus together stimulate the most growth. David W. Litke, Review of the Phosphorus Control Measures in the United States and Their Effects on Water Quality, U.S. Geological Survey, Water-Resources Investigations Report 99-4007, Denver, CO, 1999, http://pubs.usgs.gov/wri/wri994007/pdf/wri99- 4007.pdf. One example of a recent control strategy focusing on phosphorus is the International Joint Commission, A Balanced Diet for Lake Erie: Reducing Phosphorus Loadings and Harmful Algal Blooms, Report of the Lake Erie Ecosystem Priority, February 2014, http://www.ijc.org/files/publications/2014%20IJC%20LEEP%20REPORT.pdf.

[38] Centers for Disease Control and Prevention, Harmful Algal Blooms (HABs), http://www.cdc.gov/nceh/hsb/hab/ default.htm.

[39] Centers for Disease Control and Prevention, Harmful Algal Blooms (HABs), http://www.cdc.gov/nceh/hsb/hab/ default.htm.

[40] Pieter T. J. Johnson, Alan R. Townsend, and Cory C. Cleveland, et al., Linking Environmental Nutrient Enrichment and Disease Emergence in Humans and Wildlife, National Institutes of Health, public access author manuscript, April 2010.

[41] Ibid.

[42] Ward, 2008.

[43] Ibid.

[44] Ibid.

[45] Ibid. For additional information on U.S. drinking water regulations, see CRS Report RL31243, Safe Drinking Water Act (SDWA): A Summary of the Act and Its Major Requirements.

[46] USDA, NRCS, Conservation Practice Standard - Nutrient Management, Code 590, January 2012, http://www.nrcs.usda.gov/Internet/FSE_DOCUMENTS/stelprdb1046433.pdf.

[47] Also referred to as zero-till, direct (seed) drilling, slot plant, row till, strip till, or generically as conservation tillage.

[48] Noninversion tillage can include chisel plowing, field cultivating, tandem disking, or vertical tillage.

[49] Robert Mullen and Ed Lentz, Nitrogen Inhibitors, What Is What and Should You Consider Their Use? Ohio State University Extension, C.O.R.N. Newsletter 2011-08, Columbus, OH, 2011, http://corn.osu.edu/newsletters/2011/2011- 08-1/nitrogen-inhibitors-what-is-what-and-should-you-consider-their-use.

[50] While not discussed in this report, chemical fertilizers used in agriculture production are also regulated. For more information, see CRS Report R43070, Regulation of Fertilizers: Ammonium Nitrate and Anhydrous Ammonia. Also, facilities that emit large quantities of air pollutants may be regulated under the Clean Air Act. Some livestock operations also may be subject to requirements of the Comprehensive Environmental Response, Compensation, and Liability Act (CERCLA, or Superfund) and the Emergency Planning and Community Right-to-Know Act (EPCRA). For additional information, see CRS Report RL33691, Animal Waste and Hazardous Substances: Current Laws and Legislative Issues.

[51] For additional background, see CRS Report R41622, Environmental Regulation and Agriculture.

[52] For additional information, see CRS Report RL33656, Animal Waste and Water Quality: EPA's Response to the Waterkeeper Alliance Court Decision on Regulation of CAFOs.

[53] CAFOs are large animal feeding operations. The regulatory threshold of animal feeding operations that are covered by CWA regulations varies by animal type (e.g., facilities housing 700 or more mature dairy cattle; 30,000 or more laying hens or broilers), as detailed in EPA rules, which also specify CAFO pollution control requirements (40 C.F.R. Part 122 and 40 C.F.R. Part 412).

[54] U.S. Environmental Protection Agency, "Revised National Pollutant Discharge Elimination System Permit Regulation and Effluent Limitations Guidelines for Concentrated Animal Feeding Operations in Response to the Waterkeeper Decision, Final Rule," 73 Federal Register 225, November 20, 2008, pp. 70417-70486.

[55] For additional information, see CRS Report R42752, Clean Water Act and Pollutant Total Maximum Daily Loads (TMDLs).

[56] U.S. Environmental Protection Agency, National Summary of Impaired Waters and TMDL Information, http://iaspub.epa.gov/waters10/attains_nation_cy.control?p_report_type=T.

[57] Ibid.

[58] For additional information, see CRS Report R40819, USDA's Research, Education, and Economics (REE) Mission Area: Issues and Background.

[59] USDA, NRCS, Soil and Water Resources Conservation Act (RCA) Report, National Conservation Programs Profile, http://www.nrcs.usda.gov/wps/portal/nrcs/detail/national/technical/nra/rca/ida/?cid=stelprdb1187041. Nutrient management is also considered under the soil quality category of the report.

[60] Food, Conservation, and Energy Act of 2008 (P.L. 110-246). These provisions were unchanged when reauthorized in the Agricultural Act of 2014 (P.L. 113-79).

[61] For additional information on private land conservation programs, see CRS Report R40763, Agricultural Conservation: A Guide to Programs.

[62] For more information on the CWA 319 program, see http://water.epa.gov/polwaste/nps/cwact.cfm.

[63] EPA, Office of Wetlands, Oceans, and Watersheds, A National Evaluation of the Clean Water Act Section 319 Program, November 2011, http://water.epa.gov/polwaste/nps/upload/319evaluation.pdf.

[64] ARS national programs supporting nutrient related research include NP #211—Water Availability and Watershed Management, NP #212—Climate Change, Soils, and Emissions, NP #214—Agricultural and Industrial Byproducts, NP #215—Pasture, Forage and Rangeland Systems, and NP #216—Agricultural System Competitiveness and Sustainability. Other national programs, such as NP #305 (Plant Genetic Resources, Genomics and Genetic Improvement) and NP #305 (Crop Production), may also include research that impacts nutrient management through the development of plant varieties and production practices that require less nutrients. Research project information may be found at http://www.ars.usda.gov/research/programs.htm.

[65] Additional information and publication links may be found here: http://www.nrcs.usda.gov/wps/portal/nrcs/detail/ national/technical/nra/ceap/?cid=nrcs143_014153.

[66] For additional information on the National Water-Quality Assessment Program and the Agricultural Chemicals Team, see http://in.water.usgs.gov/NAWQA_ACT/.

[67] Marc Ribaudo, Jorge Delgado, and LeRoy Hansen, et al., Nitrogen in Agricultural Systems: Implications for Conservation Policy, USDA, ERS, ERR-127, Washington, DC, September 2011, p. 4.

In: Nutrient Pollution …
Editor: Arlene Walters
ISBN: 978-1-63485-187-9

Chapter 2

NUTRIENT MANAGEMENT*

Natural Resources Conservation Service

DEFINITION

Managing the amount (rate), source, placement (method of application), and timing of plant nutrients and soil amendments.

PURPOSE

- To budget, supply, and conserve nutrients for plant production.
- To minimize agricultural nonpoint source pollution of surface and groundwater resources.
- To properly utilize manure or organic byproducts as a plant nutrient source.
- To protect air quality by reducing odors, nitrogen emissions (ammonia, oxides of nitrogen), and the formation of atmospheric particulates.
- To maintain or improve the physical, chemical, and biological condition of soil.

* This is an edited, reformatted and augmented version of Conservation Practice Standard, Code 590, issued by the Natural Resources Conservation Service, January 2012.

Conditions Where Practice Applies

This practice applies to all lands where plant nutrients and soil amendments are applied. This standard does not apply to one-time nutrient applications to establish perennial crops.

Criteria

General Criteria Applicable to All Purposes

A nutrient budget for nitrogen, phosphorus, and potassium must be developed that considers all potential sources of nutrients including, but not limited to, green manures, legumes, crop residues, compost, animal manure, organic byproducts, biosolids, waste water, organic matter, soil biological activity, commercial fertilizer, and irrigation water.

Enhanced efficiency fertilizers, used in the State must be defined by the Association of American Plant Food Control Officials (AAPFCO) and be accepted for use by the State fertilizer control official, or similar authority, with responsibility for verification of product guarantees, ingredients (by AAPFCO definition) and label claims.

For nutrient risk assessment policy and procedures see Title 190, General Manual (GM), Part 402, Nutrient Management, and Title 190, National Instruction (NI), Part 302, Nutrient Management Policy Implementation.

To avoid salt damage, the rate and placement of applied nitrogen and potassium in starter fertilizer must be consistent with land-grant university guidelines, or industry practice recognized by the land-grant university.

The NRCS-approved nutrient risk assessment for *nitrogen* must be completed on all sites unless the State NRCS, with the concurrence of State water quality control authorities, has determined specific conditions where nitrogen leaching is not a risk to water quality, including drinking water.

The NRCS-approved nutrient risk assessment for *phosphorus* must be completed when:

- phosphorus application rate exceeds land-grant university fertility rate guidelines for the planned crop(s), or
- the planned area is within a phosphorus- impaired watershed (contributes to 303d-listed water bodies), or

- the NRCS and State water quality control authority have not determined specific conditions where the risk of phosphorus loss is low.

A phosphorus risk assessment will not be required when the State NRCS, with concurrence of the State water quality control authority, has determined specific conditions where the risk of phosphorus loss is low. These fields must have a documented agronomic need for phosphorus; based on soil test phosphorus (STP) and land-grant university nutrient recommendations.

On organic operations, the nutrient sources and management must be consistent with the USDA's National Organic Program.

Areas contained within minimum application setbacks (e.g., sinkholes, wellheads, gullies, ditches, or surface inlets) must receive nutrients consistent with the setback restrictions.

Applications of irrigation water must minimize the risk of nutrient loss to surface and groundwater.

Soil pH must be maintained in a range that enhances an adequate level for crop nutrient availability and utilization. Refer to State land- grant university documentation for guidance.

Soil, Manure, and Tissue Sampling and Laboratory Analyses (Testing)

Nutrient planning must be based on current soil, manure, and (where used as supplemental information) tissue test results developed in accordance with land-grant university guidance, or industry practice, if recognized by the university.

Current soil tests are those that are no older than 3 years, but may be taken on an interval recommended by the land-grant university or as required by State code. The area represented by a soil test must be that acreage recommended by the land-grant university.

Where a conservation management unit (CMU) is used as the basis for a sampling unit, all acreage in the CMU must have similar soil type, cropping history, and management practice treatment.

The soil and tissue tests must include analyses pertinent to monitoring or amending the annual nutrient budget, e.g., pH, electrical conductivity (EC) and sodicity where salts are a concern, soil organic matter, phosphorus, potassium, or other nutrients and test for nitrogen where applicable. Follow land-grant university guidelines regarding required analyses.

Soil test analyses must be performed by laboratories successfully meeting the requirements and performance standards of the North American Proficiency Testing Program-Performance Assessment Program (NAPT-PAP) under the auspices of the Soil Science Society of America (SSSA) and NRCS, or other NRCSapproved program that considers laboratory performance and proficiency to assure accuracy of soil test results. Alternate proficiency testing programs must have solid stakeholder (e.g., water quality control entity, NRCS State staff, growers, and others) support and be regional in scope.

Nutrient values of manure, organic by-products and biosolids must be determined prior to land application.

Manure analyses must include, at minimum, total nitrogen (N), ammonium N, total phosphorus (P) or P_2O_5, total potassium (K) or K_2O, and percent solids, or follow land-grant university guidance regarding required analyses.

Manure, organic by-products, and biosolids samples must be collected and analyzed at least annually, or more frequently if needed to account for operational changes (feed management, animal type, manure handling strategy, etc.) impacting manure nutrient concentrations. If no operational changes occur, less frequent manure testing is allowable where operations can document a stable level of nutrient concentrations for the preceding three consecutive years, unless federal, State, or local regulations require more frequent testing.

Samples must be collected, prepared, stored, and shipped, following land-grant university guidance or industry practice.

When planning for new or modified livestock operations, acceptable "book values" recognized by the NRCS (e.g., NRCS Agricultural Waste Management Field Handbook) and the land-grant university, or analyses from similar operations in the geographical area, may be used if they accurately estimate nutrient output from the proposed operation.

Manure testing analyses must be performed by laboratories successfully meeting the requirements and performance standards of the Manure Testing Laboratory Certification program (MTLCP) under the auspices of the Minnesota Department of Agriculture, or other NRCS- approved program that considers laboratory performance and proficiency to assure accurate manure test results.

Nutrient Application Rates

Planned nutrient application rates for nitrogen, phosphorus, and potassium must not exceed land-grant university guidelines or industry practice when recognized by the university.

At a minimum, determination of rate must be based on crop/cropping sequence, current soil test results, realistic yield goals, and NRCS- approved nutrient risk assessments.

If the land-grant university does not provide specific guidance that meets these criteria, application rates must be based on plans that consider realistic yield goals and associated plant nutrient uptake rates.

Realistic yield goals must be established based on historical yield data, soil productivity information, climatic conditions, nutrient test results, level of management, and local research results considering comparable production conditions.

Estimates of yield response must consider factors such as poor soil quality, drainage, pH, salinity, etc., prior to assuming that nitrogen and/or phosphorus are deficient.

For new crops or varieties, industry- demonstrated yield, and nutrient utilization information may be used until land-grant university information is available.

Lower-than-recommended nutrient application rates are permissible if the grower's objectives are met.

Applications of biosolids, starter fertilizers, or pop-up fertilizers must be accounted for in the nutrient budget.

Nutrient Sources

Nutrient sources utilized must be compatible with the application timing, tillage and planting system, soil properties, crop, crop rotation, soil organic content, and local climate to minimize risk to the environment.

Nutrient Application Timing and Placement

Timing and placement of all nutrients must correspond as closely as practical with plant nutrient uptake (utilization by crops), and consider nutrient source, cropping system limitations, soil properties, weather conditions, drainage system, soil biology, and nutrient risk assessment results.

Nutrients must not be surface-applied if nutrient losses offsite are likely. This precludes spreading on:

- frozen and/or snow-covered soils, and
- when the top 2 inches of soil are saturated from rainfall or snow melt.

Exceptions for the above criteria can be made for surface-applied manure when specified conditions are met and adequate conservation measures are installed to prevent the offsite delivery of nutrients. The adequate treatment level and specified conditions for winter applications of manure must be defined by NRCS in concurrence with the water quality control authority in the State. At a minimum, the following site and management factors must be considered:

- slope,
- organic residue and living covers,
- amount and form of nutrients to be applied, and
- adequate setback distances to protect local water quality.

Additional Criteria to Minimize Agricultural Nonpoint Source Pollution of Surface and Groundwater

Planners must use the current NRCS-approved nitrogen, phosphorus, and soil erosion risk assessment tools to assess the risk of nutrient and soil loss. Identified resource concerns must be addressed to meet current planning criteria (quality criteria). Technical criteria for risk assessments can be found in NI-190-302.

When there is a high risk of transport of nutrients, conservation practices must be coordinated to avoid, control, or trap manure and nutrients before they can leave the field by surface or subsurface drainage (e.g., tile). The number of applications and the application rates must also be considered to limit the transport of nutrients to tile.

Nutrients must be applied with the right placement, in the right amount, at the right time, and from the right source to minimize nutrient losses to surface and groundwater. The following nutrient use efficiency strategies or technologies must be considered:

- slow and controlled release fertilizers
- nitrification and urease inhibitors
- enhanced efficiency fertilizers
- incorporation or injection
- timing and number of applications
- soil nitrate and organic N testing
- coordinate nutrient applications with optimum crop nutrient uptake
- Corn Stalk Nitrate Test (CSNT), PreSidedress Nitrate Test (PSNT), and Pre-Plant Soil Nitrate Test (PPSN)
- tissue testing, chlorophyll meters, and spectral analysis technologies
- other land-grant university recommended technologies that improve nutrient use efficiency and minimize surface or groundwater resource concerns.

Additional Criteria Applicable to Properly Utilize Manure or Organic By-Products as a Plant Nutrient Source

When manures are applied, and soil salinity is a concern, salt concentrations must be monitored to prevent potential crop damage and/or reduced soil quality.

The total single application of liquid manure:

- must not exceed the soil's infiltration or water holding capacity
- be based on crop rooting depth
- must be adjusted to avoid runoff or loss to subsurface tile drains.

Crop production activities and nutrient use efficiency technologies must be coordinated to take advantage of mineralized plant-available nitrogen to minimize the potential for nitrogen losses due to denitrification or ammonia volatilization.

Nitrogen and phosphorus application rates must be planned based on risk assessment results as determined by NRCS-approved nitrogen and phosphorus risk assessment tools.

For fields receiving manure, where phosphorus risk assessment results equate to LOW risk, additional phosphorus and potassium can be applied at rates greater than crop requirement not to exceed the nitrogen requirement for the succeeding crop. For fields receiving manure, where phosphorus risk assessment results equate to MODERATE risk, additional phosphorus and potassium may be applied at a phosphorus crop requirement rate for the planned crops in the rotation. When phosphorus risk assessment results equate to HIGH risk, additional phosphorus and potassium may be applied at phosphorus crop removal rates if the following requirements are met:

- a soil phosphorus drawdown strategy has been implemented, and
- a site assessment for nutrients and soil loss has been conducted to determine if mitigation practices are required to protect water quality.
- any deviation from these high risk requirements must have the approval of the Chief of the NRCS.

Manure or organic by-products may be applied on legumes at rates equal to the estimated removal of nitrogen in harvested plant biomass, not to exceed land grant university recommendations.

Manure may be applied at a rate equal to the recommended phosphorus application, or estimated phosphorus removal in harvested plant biomass for the crop rotation, or multiple years in the crop sequence at one time. When such applications are made, the application rate must not exceed the acceptable phosphorus risk assessment criteria, must not exceed the recommended nitrogen application rate during the year of application or harvest cycle, and no additional phosphorus must be applied in the current year and any additional years for which the single application of phosphorus is supplying nutrients.

Additional Criteria to Protect Air Quality by Reducing Odors, Nitrogen Emissions and the Formation of Atmospheric Particulates

To address air quality concerns caused by odor, nitrogen, sulfur, and/or particulate emissions; the source, timing, amount, and placement of nutrients must be adjusted to minimize the negative impact of these emissions on the environment and human health. One or more of the following may be used:

- slow or controlled release fertilizers
- nitrification inhibitors
- urease inhibitors
- nutrient enhancement technologies
- incorporation
- injection
- stabilized nitrogen fertilizers
- residue and tillage management
- no-till or strip-till
- other technologies that minimize the impact of these emissions

Do not apply poultry litter, manure, or organic by-products of similar dryness/density when there is a high probability that wind will blow the material offsite.

ADDITIONAL CRITERIA TO IMPROVE OR MAINTAIN THE PHYSICAL, CHEMICAL, AND BIOLOGICAL CONDITION OF THE SOIL TO ENHANCE SOIL QUALITY FOR CROP PRODUCTION AND ENVIRONMENTAL PROTECTION

Time the application of nutrients to avoid periods when field activities will result in soil compaction.

In areas where salinity is a concern, select nutrient sources that minimize the buildup of soil salts.

Considerations

Elevated soil test phosphorus levels are detrimental to soil biota. Soil test phosphorus levels should not exceed State-approved soil test thresholds established to protect the environment.

Use no-till/strip-till in combination with cover crops to sequester nutrients, increase soil organic matter, increase aggregate stability, reduce compaction, improve infiltration, and enhance soil biological activity to improve nutrient use efficiency.

Use nutrient management strategies such as cover crops, crop rotations, and crop rotations with perennials to improve nutrient cycling and reduce energy inputs.

Use variable-rate nitrogen application based on expected crop yields, soil variability, soil nitrate or organic N supply levels, or chlorophyll concentration.

Use variable-rate nitrogen, phosphorus, and potassium application rates based on site-specific variability in crop yield, soil characteristics, soil test values, and other soil productivity factors.

Develop site-specific yield maps using a yield monitoring system. Use the data to further diagnose low- and high- yield areas, or zones, and make the necessary management changes. See Title 190, Agronomy Technical Note (TN) 190.AGR.3, Precision Nutrient Management Planning.

Use manure management conservation practices to manage manure nutrients to limit losses prior to nutrient utilization.

Apply manure at a rate that will result in an "improving" Soil Conditioning Index (SCI) without exceeding acceptable risk of nitrogen or phosphorus loss.

Use legume crops and cover crops to provide nitrogen through biological fixation and nutrient recycling.

Modify animal feed diets to reduce the nutrient content of manure following guidance contained in Conservation Practice Standard (CPS) Code 592, Feed Management.

Soil test information should be no older than 1 year when developing new plans.

Excessive levels of some nutrients can cause induced deficiencies of other nutrients, e.g., high soil test phosphorus levels can result in zinc deficiency in corn.

Use soil tests, plant tissue analyses, and field observations to check for secondary plant nutrient deficiencies or toxicity that may impact plant growth or availability of the primary nutrients.

Use the adaptive nutrient management learning process to improve nutrient use efficiency on farms as outlined in the NRCS' National Nutrient Policy in GM 190, Part 402, Nutrient Management.

Potassium should not be applied in situations where an excess (greater than soil test potassium recommendation) causes nutrient imbalances in crops or forages.

Workers should be protected from and avoid unnecessary contact with plant nutrient sources. Extra caution must be taken when handling anhydrous ammonia or when dealing with organic wastes stored in unventilated enclosures.

Material generated from cleaning nutrient application equipment should be utilized in an environmentally safe manner. Excess material should be collected and stored or field applied in an appropriate manner.

Nutrient containers should be recycled in compliance with State and local guidelines or regulations.

CONSIDERATIONS TO MINIMIZE AGRICULTURAL NONPOINT SOURCE POLLUTION OF SURFACE AND GROUNDWATER

Use conservation practices that slow runoff, reduce erosion, and increase infiltration, e.g., filter strip, contour farming, or contour buffer strips. These practices can also reduce the loss of nitrates or soluble phosphorus.

Use application methods and timing strategies that reduce the risk of nutrient transport by ground and surface waters, such as:

- split applications of nitrogen to deliver nutrients during periods of maximum crop utilization,
- banded applications of nitrogen and/or phosphorus to improve nutrient availability,
- drainage water management to reduce nutrient discharge through drainage systems, and
- incorporation of surface-applied manures or organic by-products if precipitation capable of producing runoff or erosion is forecast within the time of planned application.

Use the agricultural chemical storage facility conservation practice to protect air, soil, and water quality.

Use bioreactors and multistage drainage strategies when approved by the land-grant university.

Considerations to Protect Air Quality by Reducing Nitrogen and/or Particulate Emissions to the Atmosphere

Avoid applying manure and other by-products upwind of inhabited areas.

Use high-efficiency irrigation technologies (e.g., reduced-pressure drop nozzles for center pivots) to reduce the potential for nutrient losses.

Plans and Specifications

The following components must be included in the nutrient management plan:

- aerial site photograph(s)/imagery or site map(s), and a soil survey map of the site,
- soil information including: soil type surface texture, pH, drainage class, permeability, available water capacity, depth to water table, restrictive features, and flooding and/or ponding frequency,
- location of designated sensitive areas and the associated nutrient application restrictions and setbacks,
- for manure applications, location of nearby residences, or other locations where humans may be present on a regular basis, and any identified meteorological (e.g., prevailing winds at different times of the year), or topographical influences that may affect the transport of odors to those locations,
- results of approved risk assessment tools for nitrogen, phosphorus, and erosion losses,
- documentation establishing that the application site presents low risk for phosphorus transport to local water when phosphorus is applied in excess of crop requirement.
- current and/or planned plant production sequence or crop rotation,
- soil, water, compost, manure, organic byproduct, and plant tissue sample analyses applicable to the plan,
- when soil phosphorus levels are increasing, include a discussion of the risk associated with phosphorus accumulation and a proposed phosphorus draw-down strategy,
- realistic yield goals for the crops,

- complete nutrient budget for nitrogen, phosphorus, and potassium for the plant production sequence or crop rotation,
- listing and quantification of all nutrient sources and form,
- all enhanced efficiency fertilizer products that are planned for use,
- in accordance with the nitrogen and phosphorus risk assessment tool(s), specify the recommended nutrient application source, timing, amount (except for precision/variable rate applications specify method used to determine rate), and placement of plant nutrients for each field or management unit, and
- guidance for implementation, operation and maintenance, and recordkeeping.

In addition, the following components must be included in a precision/variable rate nutrient management plan:

- Document the geo-referenced field boundary and data collected that was processed and analyzed as a GIS layer or layers to generate nutrient or soil amendment recommendations.
- Document the nutrient recommendation guidance and recommendation equations used to convert the GIS base data layer or layers to a nutrient source material recommendation GIS layer or layers.
- Document if a variable rate nutrient or soil amendment application was made.
- Provide application records per management zone or as applied map within individual field boundaries (or electronic records) documenting source, timing, method, and rate of all applications that resulted from use of the precision agriculture process for nutrient or soil amendment applications.
- Maintain the electronic records of the GIS data layers and nutrient applications for at least 5 years.

If increases in soil phosphorus levels are expected (i.e., when N-based rates are used), the nutrient management plan must document:

- the soil phosphorus levels at which it is desirable to convert to phosphorus based planning,
- the potential plan for soil test phosphorus drawdown from the production and harvesting of crops, and

- management activities or techniques used to reduce the potential for phosphorus transport and loss,
- for AFOs, a quantification of manure produced in excess of crop nutrient requirements, and
- a long-term strategy and proposed implementation timeline for reducing soil P to levels that protect water quality

Operation and Maintenance

Conduct periodic plan reviews to determine if adjustments or modifications to the plan are needed. At a minimum, plans must be reviewed and revised, as needed with each soil test cycle, changes in manure volume or analysis, crops, or crop management.

Fields receiving animal manures and/or biosolids must be monitored for the accumulation of heavy metals and phosphorus in accordance with land-grant university guidance and State law.

Significant changes in animal numbers, management, and feed management will necessitate additional manure analyses to establish a revised average nutrient content.

Calibrate application equipment to ensure accurate distribution of material at planned rates.

Document the nutrient application rate. When the applied rate differs from the planned rate, provide appropriate documentation for the change.

Records must be maintained for at least 5 years to document plan implementation and maintenance. As applicable, records include:

- soil, plant tissue, water, manure, and organic by-product analyses resulting in recommendations for nutrient application,
- quantities, analyses and sources of nutrients applied,
- dates, and method(s) of nutrient applications, source of nutrients, and rates of application,
- weather conditions and soil moisture at the time of application; lapsed time to manure incorporation; rainfall or irrigation event,
- crops planted, planting and harvest dates, yields, nutrient analyses of harvested biomass, and crop residues removed,
- dates of plan review, name of reviewer, and recommended changes resulting from the review, and
- all enhanced efficiency fertilizer products used.

Additional records for precision/variable rate sites must include:

- maps identifying the variable application source, timing, amount, and placement of all plant nutrients applied, and
- GPS-based yield maps for crops where yields can be digitally collected.

REFERENCES

Association of American Plant Food Control Officials (AAPFCO). 2011. AAPFCO Official Publication no. 64. AAPFCO Inc., Little Rock, AR.

Follett, R.F. 2001. Nitrogen transformation and transport processes. *In* Nitrogen in the environment; sources, problems, and solutions, (eds.) R.F. Follett and J. Hatfield, pp. 17-44. Elsevier Science Publishers. The Netherlands. 520 pp.

Schepers, J.S., and W.R. Ruan, (eds.) 2008. Nitrogen in agricultural systems. Agron. Monogr. no. 49, American Society of Agronomy (ASA), Crop Science Society of America (CSSA), Soil Science Society of America (SSSA). Madison, WI.

Sims, J.T. (ed.) 2005. Phosphorus: Agriculture and the environment. Agron. Monogr. no. 46. ASA, CSSA, and SSSA, Madison, WI.

Stevenson, F.J. (ed.) 1982. Nitrogen in agricultural soils. Agron. Series 22. ASA, CSSA, and SSSA, Madison, WI.

U.S. Department of Agriculture, Natural Resources Conservation Service. 2010. Agronomy Technical Note, (TN) 190-AGR-3, Precision Nutrient Management Planning. Washington, DC.

U.S. Department of Agriculture, Natural Resources Conservation Service. 2011. Title 190, General Manual, (GM), Part 402, Nutrient Management. Washington, DC.

U.S. Department of Agriculture, Natural Resources Conservation Service. 2011, Title 190, National Instruction (NI), Part 302, Nutrient Management Policy Implementation. Washington, DC.

In: Nutrient Pollution ...
Editor: Arlene Walters

ISBN: 978-1-63485-187-9

Chapter 3

AN ECONOMIC ASSESSMENT OF POLICY OPTIONS TO REDUCE AGRICULTURAL POLLUTANTS IN THE CHESAPEAKE BAY*

Marc Ribaudo, Jeffrey Savage and Marcel Aillery

ABSTRACT

In 2010, a Total Maximum Daily Load (TMDL) was established for the Chesapeake Bay, defining the limits on emissions of nitrogen, phosphorus, and sediment necessary to reverse declines in the Bay's quality and associated biological resources. Agriculture is the largest single source of nutrients and sediment in the watershed. We use data on crop and animal agriculture in the watershed to assess the relative effectiveness of alternative policy approaches for achieving the nutrient and sediment reduction goals of the TMDL, ranging from voluntary financial incentives to regulations. The cost of achieving water quality goals depends heavily on which policy choices are selected and how they are implemented. We found that policies that provide incentives for water quality improvements are the most efficient, assuming necessary information on pollutant delivery is available for each field. Policies that directly encourage adoption of management systems that protect water quality (referred to as design-based) are the most practical, given the limited information that is generally available to farmers and resource

---

* This is an edited, reformatted and augmented version of Economic Research Report Number 166, issued by the U.S. Department of Agriculture, Economic Research Service, June 2014.

agencies. Information on field characteristics can be used to target design-based policies to improve efficiency.

Keywords: Chesapeake Bay, sediment, manure, water quality, economic incentives, conservation policy, regulation, phosphorus, nitrogen, TMDL

SUMMARY

What Is the Issue?

In 2010, the U.S. Environmental Protection Agency established limits for nutrient and sediment emissions from point (i.e., wastewater treatment plant) and nonpoint (i.e., agricultural runoff) sources to the Chesapeake Bay in the form of a Total Maximum Daily Load (TMDL). Agriculture is the largest single source of nutrient emissions in the watershed. The TMDL specifies that two issues facing agriculture need to be addressed if the TMDL's nutrient limits are to be met. First, farmers can increase their use of the most effective nutrient management practices, such as cover crops. Second, assuming animal numbers in the watershed do not change, reducing the amount of manure applied to farmland necessitates moving manure to areas with cropland that can safely receive manure nutrients as a substitute for commercial fertilizer. Both entail costs.

Different policy approaches can be used to achieve the nutrient limits of the TMDL, ranging from financial incentives to regulations. The cost of achieving water quality goals depends heavily on which policy choices are selected and how they are implemented.

What Did the Study Find?

Our assessment of policy instruments for achieving the TMDL goals on cropland in the Chesapeake Bay watershed includes performance-based regulations (emission limits), performance-based incentives (emission taxes), design-based standards (regulations on practices or inputs), and design-based incentives (payments for conservation practices). Our analysis focuses on the difference in costs between scenarios, rather than the absolute cost of any one policy scenario. These differences enable us to identify which policy design features lead to a more cost-effective solution.

Performance-based policies use measures of pollutant delivered to receiving waters as the basis for policy. The lowest cost ("optimal") option among all approaches meets TMDL goals by treating only 12 percent of the watershed's cropland—those acres that can reduce the most pollutants at the least cost. Performance-based policies are difficult to implement for nonpoint-source pollution because pollutant discharge cannot be easily measured and regulators lack the information necessary to set optimal performance goals.

An alternative is to focus on inputs and management practices, which are more easily observed by both resource management agencies and farmers. Practices can be required through regulation or encouraged through financial incentives. These ***design-based policies*** can be made more efficient through targeting. In particular, requiring best management practices—a combination of cover crops, nutrient management, and erosion controls—on land adjacent to water that also has a treatment need according to USDA's Natural Resources Conservation Service (NRCS) criteria met TMDL goals for a quarter of the cost of implementing the full suite of management practices on all cropland. Still, the most efficient design-based approach was 4-5 times more expensive than the optimal performance-based approach and treated more than twice as much land.

Another incentive system for Bay-area farmers to reduce their pollutant levels is the establishment of ***water quality trading***. Trading is already used by some States to enable wastewater treatment plants and other point sources to meet TMDL limits without costly upgrades. Trading also offers an economic incentive to farmers to implement water quality-improving practices they might not otherwise adopt. Trading program rules developed by Maryland, Pennsylvania, and Virginia all require that fields achieve a certain level of nutrient management before being able to generate credits, thus ensuring that abatement contributed by agriculture is "additional." However, requiring more of farmers in order to trade can discourage them from participating.

With a daily inventory of roughly 2.0 billion pounds of poultry, dairy, swine, and feedlot beef, ***animal agriculture is a significant source of nutrient loadings*** to the Bay, contributing an estimated 17 percent of the nitrogen entering the Bay and 26 percent of the phosphorus. Animal operations produce roughly 99,400 tons of recoverable manure nitrogen and 44,200 tons of recoverable manure phosphorus annually, which is often more than can be safely used by crops grown on the land managed by the livestock operation. Removing excess manure nutrients from farms in the Chesapeake Bay watershed and moving it to where it can be used efficiently costs an estimated $15 million to $27 million per year.

Manure hauling costs decrease with the ***willingness of crop producers to use manure***. An increase in the share of cropland using manure from 30 to 90 percent reduces hauling costs in the watershed about 15 percent. Education and technical/financial assistance for manure management could increase the willingness of crop producers to substitute manure for commercial fertilizers.

Using manure as an energy source could absorb some of the excess manure nutrients and reduce regional hauling costs if the economics are favorable and concerns over air quality are adequately addressed. However, demand for manure as an energy source could increase costs for crop producers who utilize manure as a nutrient source, either through higher manure prices or through the need to purchase inorganic fertilizers. Our findings suggest that the value of manure as a source of crop nutrients would exceed the reduction in hauling costs. But this negative impact on agriculture would have to be weighed against the benefits provided by local energy production.

How Was the Study Conducted?

To evaluate the costs of improved nutrient management on cropland, the study used data from NRCS's Conservation Effects Assessment Project (CEAP). Cropping practice data, simulated nutrient emissions to water (also from CEAP), and practice cost data from a number of sources were used to build a model to evaluate different policy scenarios for meeting TMDL goals, subject to policy constraints.

The CEAP data and the optimization model were also used to examine the implications of baseline choice in a point/nonpoint trading program. The data and model were used to estimate nitrogen abatement supply curves for different baseline assumptions. An offset demand curve for publicly owned wastewater treatment plants was estimated by the World Resources Institute using data from the Chesapeake Bay Program. To evaluate the cost of achieving a manure nutrient balance in the watershed, the study used an ERS optimization model of the Chesapeake Bay watershed that minimizes manure hauling and application costs and is based on a county-level dataset of manure nutrients and available cropland that NRCS developed from the 2007 Census of Agriculture.

INTRODUCTION

The Chesapeake Bay is North America's largest and most biologically diverse estuary. Its watershed covers 64,000 square miles across 6 States (Delaware, Maryland, New York, Pennsylvania, Virginia, and West Virginia) and the District of Columbia and is home to more than 17 million people. For over 200 years, the Chesapeake Bay has provided a rich bounty of crabs, shellfish, and fish, as well as high-quality recreational opportunities. However, as the region's population grew and land was converted from forests to farms and urban development, the quality of the Bay's waters declined along with many of its living resources. Since the 1970s, the Clean Water Act has facilitated significant reductions in polluting emissions from sewage treatment plants, factories, and other point sources of pollution in the Bay watershed, but point sources are only part of the problem in meeting established water quality goals. In particular, continued heavy nutrient and sediment runoff from nonpoint sources has contributed to low oxygen levels, algal blooms, decreased water clarity, loss of submerged aquatic vegetation, and declines in fish and shellfish populations (U.S. EPA, 2011). In 2010, only 18 percent of tidal waters met or exceeded guidelines for water clarity; only 38 percent of the Bay and its tidal tributaries met Clean Water Act standards for dissolved oxygen; and more than half of stream health scores at monitoring sites were poor (Chesapeake Bay Program, 2011).

The history of efforts to restore the ecosystem of the Chesapeake Bay is emblematic of a general failure to solve the agricultural nonpoint-source (NPS) problem. The Bay has been a focal point of Federal and State initiatives to reduce nutrient and sediment pollution from agriculture and other sources for almost 30 years (see box, "History of Chesapeake Bay Restoration Efforts"). Beginning in 1983, successive agreements between the U.S. Environmental Protection Agency (EPA); the Governors of Maryland, Pennsylvania, Virginia, West Virginia, Delaware, and New York; and the Mayor of the District of Columbia established the Chesapeake Bay Program, set nutrient and sediment reduction goals, and developed strategies that rely heavily on voluntary approaches for reducing nutrient and sediment emissions. The U.S. Department of Agriculture (USDA)—along with EPA and the watershed States—has made large public investments to improve the management of agricultural resources and to reduce agriculture's negative impact on environmental quality. Monitoring data and model estimates based on land use changes observed between 1985 and 2009 indicate some reductions in nutrient and sediment loads (National Research Council, 2011). Despite this progress, the biological health of the Bay remains poor (CBP, 2011).

In 2009, President Barack Obama issued the Chesapeake Bay Protection and Restoration Executive Order, which calls for the Federal Government to lead a renewed effort to restore and protect the Bay and its watershed. In 2010, the U.S. EPA established a Total Maximum Daily Load for the Bay (see box, "The Total Maximum Daily Load"). It sets emission limits for nitrogen, phosphorus, and sediment across the Bay jurisdictions that are believed necessary to meet applicable water quality standards in the Bay and its tidal rivers and embayments.

History of Chesapeake Bay Restoration Efforts

In 1976, Congress directed the U.S. Environmental Protection Agency (EPA) to undertake a comprehensive study of the Bay's condition and what measures would be necessary to restore it to its former health. That effort culminated in the 1983 report, *Chesapeake Bay: A Framework for Action*, which described the state of the Bay's ecosystem, its change over time, and scientific evaluations of the Bay's functions in relation to its condition (U.S. EPA, 1983). The report established a regional management framework for addressing some of the Bay's most significant problems. Among the recommendations made in the report was for Bay watershed States to take action to reduce pollution—primarily excess nutrients and sediment—from agricultural and urban sources.

The first Bay agreement in 1983—between Maryland, Virginia, Pennsylvania, the District of Columbia, and the U.S. EPA—simply acknowledged there was a problem and called for the development of plans to address excess nutrient and sediment pollution (CBP, 1983). In 1987, a new agreement set a goal of reducing controllable nitrogen and phosphorus loads delivered to the Bay by 40 percent by 2000 (CBP, 1987). A 1991 evaluation led to an increased focus on Bay tributaries and added the program goal of expanding the total area of submerged aquatic vegetation (CBP, 1992).

An evaluation in 1997 found some progress toward reducing nitrogen and phosphorus, but concluded that restoration efforts needed to accelerate in order to both maintain what had been achieved and to meet the 2000 goals (CBP, 1999). In 2000, a new agreement—joined by Delaware, New York, and eventually West Virginia—reaffirmed the 40-percent reduction goal and targeted 2010 for progress sufficient to remove the Bay and tributary waters from the U.S. EPA list of impaired waters (CBP, 2000).

By 2006, the Bay jurisdictions had developed Tributary Strategies that outlined river basin-specific implementation activities for reducing nitrogen, phosphorus, and sediment loads from point and nonpoint sources. Implementation strategies for nonpoint sources relied almost exclusively on voluntary approaches. A 2007 evaluation concluded that insufficient progress was being made toward load reductions (CBP, 2007), prompting President Obama's Executive Order (13508) in 2009, which calls for the Federal Government to lead a renewed effort to restore and protect the Bay and its watershed. The Chesapeake Bay TMDL was finalized in 2010.

The Chesapeake Bay TMDL is the largest ever developed by the U.S. EPA, in terms of geographic coverage. Meeting the emission limits requires basinwide reductions in nitrogen, phosphorus, and sediment of 25 percent, 24 percent, and 20 percent from 2009 loads, respectively (U.S. EPA, 2010b). States are to have in place all pollution control measures needed to meet the TMDL by 2025. Because of agriculture's significant share of pollutant contributions, meeting the TMDL goals will not be possible without changes in the way agriculture is conducted in the watershed. The TMDL is forcing States to consider expanding the set of policy approaches that have been traditionally applied to agricultural nonpoint-source pollution.

The Bay Total Maximum Daily Load

A Total Maximum Daily Load (TMDL) is a calculation of the maximum amount of a pollutant that a water body can receive and still meet water quality standards. TMDLs are intended to be the common basis for control of point and nonpoint sources of pollution for surface waters that are designated as impaired under Section 303(d) of the Clean Water Act.

The TMDL provisions of the Clean Water Act (CWA) are designed as a second line of defense for protecting the quality of surface-water resources. Section 402 of the CWA established the National Pollutant Discharge Elimination System (NPDES) to restrict the discharge of pollutants from municipal and industrial dischargers. Each point source must obtain a discharge permit before it can discharge wastes into surface water. The permit requires dischargers to comply with technology-based controls (uniform, U.S. EPA-established standards of treatment that apply to certain industries and municipal sewage treatment facilities) or water quality-based controls that invoke State numeric or narrative (qualitative) water quality standards.

Section 305(b) of the CWA requires States to report to Congress every 2 years which waters are meeting water quality standards and which are not meeting the standards. When technology-based controls are inadequate for waters to meet State water quality standards, Section 303(d) of the CWA requires States to submit to the U.S. EPA a list of impaired waters and the cause of the impairment. Once listed as impaired, a TMDL may be required, along with a plan to mitigate the impairment and meet the applicable water quality standard.

Developing a TMDL involves calculating the maximum amount of a pollutant that a water body can receive and still meet water quality standards, and allocating pollutant loads to the various sources. The TMDL for the watershed is the sum of individual wasteload allocations for point sources, load allocations for nonpoint sources and natural sources (i.e., runoff from undisturbed forests and other natural areas), and a margin of safety. Waste load allocations for point sources are met through the NPDES permit, which is regulatory in nature. Load allocations for nonpoint sources can be met through regulatory or voluntary approaches developed by the States. Since the Clean Water Act specifically bars the use of Federal regulations to reduce nonpoint-source pollution, regulations would have to originate at the State level.

Current Status of Policies for Reducing Agricultural Impact on Water Quality

Agriculture is the largest contributor of nutrients and sediment to the Bay. Crop production and animal operations contributed about 38 percent of total nitrogen loads, 45 percent of total phosphorus loads, and 60 percent of total sediment loads in 2007 (National Academy of Sciences, 2011). The Clean Water Act is the main Federal law for addressing water pollution. However, nonpoint sources such as crop production are exempt from the permitting requirements of the Act. The burden of pollution reduction is on point sources, which are regulated by the National Point Discharge Elimination System (NPDES). U.S. EPA estimates that only 49 percent of nitrogen, 35 percent of phosphorus, and 4 percent of sediment loads entering the Bay are subject to Federal regulation (U.S. EPA, 2010a). The remainder must be addressed through non-regulatory provisions of the Clean Water Act, voluntary conservation programs administered by USDA, and State policies.

Prior to the TMDL, the Bay States relied heavily on traditional USDA conservation programs, which use financial incentives to encourage farmers to voluntarily adopt management measures consistent with the goals laid out by the various Chesapeake Bay agreements. These conservation programs include the Environmental Quality Incentives Program (EQIP), the Conservation Reserve Program (CRP), and the Conservation Reserve Enhancement Program (CREP). Watershed Implementation Plans (WIPs), developed by the Bay Watershed States outlining their strategies for meeting the TMDL, show a continued reliance on Federal and State financial assistance for practices such as cover crops on cropland. However, exclusive reliance on such policies has not been effective in improving water quality (Shortle et al., 2012). Despite years of conservation expenditures, nearly 90 percent of cropland in the Bay watershed is in need of additional nutrient or sediment management measures (USDA, NRCS, 2011). High expected costs and farmer resistance have hindered the use of practices necessary to reduce agricultural runoff (Bosch et al., 1992; Osmond et al., 2012).

All of the Bay States have developed or are developing water quality trading programs. Trading programs provide regulatory flexibility for point sources, allowing them to purchase abatement from other sources, including nonpoint sources, that exceed a compliance baseline. Ideally, trading reduces the cost of meeting TMDL goals and increases abatement by nonpoint sources. Each State has its own set of rules for establishing eligibility (baselines) and calculating credits or offsets. Program design plays a critical role in the attractiveness of trading to potential program participants and their ability to meet their TMDL goals.

In this report, we focus on two issues. Not enough cropland is under the most effective (best) management practices for protecting water quality (USDA, NRCS, 2011; U.S. EPA, USDA, 2006; NRC, 2011; Lichtenberg et al., 2010), and the State WIPs seek to redress that. A model built with data compiled specifically for the Chesapeake Bay watershed enables us to evaluate the efficiency characteristics of different policy instruments for achieving TMDL goals for nutrient and sediment emissions. The policy instruments examined span those contained in the State WIPs as well as alternatives, and include regulation, taxes, financial incentives, and water quality trading.

Animal operations in the Bay watershed generally produce more manure nutrients than can be assimilated by the crops grown on the operation, and addressing this imbalance is another goal of the State WIPs. Farm-level implementation of improved nutrient management practices on cropland cannot be effective on these operations until the problem of excess manure production is addressed. A model built with data on animal operations in the watershed helps to evaluate the cost of balancing manure production and manure-nutrient demand by redistributing manure around the watershed.

POLICY INSTRUMENTS FOR REDUCING AGRICULTURAL POLLUTION

Farmers operate in a profit-seeking framework. They consider input and output prices and their resource base when making decisions about what mix of crops to grow and how to grow them. In this simple framework, the cost of nutrient or sediment runoff of nutrients or sediment from fields is not considered, as farmers typically do not bear the downstream costs these pollutants may impose. The goal of nonpoint-source policy is to induce farmers who cause water quality damages through their production and management decisions to adopt pollution prevention and control measures that are consistent with societal goals (in this case defined by the TMDL). Policymakers can consider a range of policy instruments to accomplish this, from purely voluntary to regulatory. Selecting a policy approach for reducing agricultural nonpoint-source pollution requires making several choices: whom to target, what to target, and what stimulus to use.

Whom to Target

Cost-effective control of nonpoint-source pollution acknowledges the heterogeneity of agriculture by promoting different environmental performance goals for different farms so as to minimize costs over the landscape (also known as allocative efficiency). A long line of research has found that certain cropland—defined in terms of resource characteristics, farming practices, and geographic location—tends to contribute a disproportionate share of pollutants degrading a water resource (Ribaudo, 1986; 1989; Veith et al., 2004; Diebel et al., 2008). A policy promotes cost-effectiveness when it adequately accounts for the variation in each farm's water quality impacts and abatement costs (Babcock et al., 1997; Braden et al., 1989; Claassen and Horan, 2001; Fleming and Adams, 1997). For example, in the Chesapeake Bay watershed, onsite losses of nutrients and sediment differ widely by watershed, defined here as a 4-digit HUC[1] (table 1) (USDA, NRCS, 2011). Pre-TMDL per-acre losses of nitrogen, phosphorus, and sediment are highest in the Susquehanna watershed (HUC 205), followed by the Potomac watershed (HUC 207).

However, assigning responsibility for nonpoint source pollution loads is difficult. Runoff does not emanate from a single point, but leaves each field in so many places, often unseen, that accurate monitoring would be prohibitively expensive (Braden and Segerson, 1993; Shortle and Abler, 1997). The amount and nature of runoff leaving a field depend not only on factors that can be measured, such as the technology employed and the use of variable inputs, but also on factors such as rainfall that vary daily and are difficult to predict (Braden and Segerson, 1993; Shortle and Abler, 1997).

Uncertainty about who is responsible and the degree of responsibility creates problems for efficient nonpoint-source policy design. Requiring farmers who are not contributing to the problem to adopt best management practices reduces program cost-effectiveness. Similarly, not covering those farmers who can provide the greatest reductions at the lowest cost also reduces cost-effectiveness. The difficulty in identifying which producers are producing the most pollution is an often-cited reason why voluntary actions are the tools of choice in existing nonpoint-source programs (Shortle and Horan, 2001).

Table 1. Average annual baseline losses at edge of field, by sub-basin (4-digit Hydrologic Unit Code (HUC))

Loss pathway	HUC 205 Susquehanna	HUC 206 Upper Chesapeake	HUC 207 Potomac	HUC 208 Lower Chesapeake	Basin
Nitrogen loss (lbs./acre)					
Surface runoff	12.9	5.4	8.3	4.9	8.9
Subsurface	43.2	25.7	32.5	18.2	32.7
Total	56.1	31.1	40.8	33.1	41.6
Phosphorus loss (lbs./acre)					
Surface runoff	5.25	2.07	4.39	1.67	3.67
Subsurface	.08	.06	.07	.06	.07
Total	5.33	2.13	4.46	1.73	3.74
Sediment (tons/acre)	2.05	0.43	1.02	0.43	1.18

Source: USDA, Natural Resources Conservation Service, 2011

What to Target

Since the goal of a nonpoint-source policy for the Chesapeake Bay watershed is to improve water quality in the Bay, the ideal policy would base incentives (positive or negative) on pollutant delivery to tidal waters (Oates, 1995). However, measuring nonpoint-source emissions is extremely difficult and costly, making actual emissions delivery impractical to use in a policy. Alternative instruments should be highly correlated with environmental conditions, enforceable, and targetable in time and space (Braden and Segerson, 1993).

One option is using estimated loadings within the tidal waters of the Bay. The development of accepted field- and watershed-level modeling tools allows estimated emissions to be used as a policy base. For example, estimates of soil erosion from the Universal Soil Loss Equation are being used in USDA conservation compliance policy (Code of Federal Regulations, 1996). Emissions estimates from models such as the Nutrient Tracking Tool developed by USDA's Natural Resources Conservation Service (NRCS) and the Agricultural Policy Environmental Extender (APEX) could be used as a basis for policies addressing nutrient pollution. Shortcomings of this approach include the need to run the model for each farm and multiple farming options, which increases transaction costs. Also, estimates of runoff are available only for the management choices contained in the model. And the degree of uncertainty surrounding model estimates would need to be accounted for in setting policy.

Other policy design options are to directly influence the inputs or farming practices that are correlated with pollutant flows. These may include inputs such as fertilizer, pesticides, or irrigation water, and practices such as conservation tillage and nutrient management. Farming practices are generally easier to observe. The drawback of input- or design-based options is that they do not account for factors between the field and the receiving waters (i.e. slope and vegetation) that influence water pollution in the Bay (the policy goal). Policies designed around factors that are more correlated with emissions are more efficient (Ribaudo et al., 1999).

What Type of Incentive

The policy stimulus is the incentive the producer responds to when making management choices. The least intrusive instruments are public persuasion combined with education and technical assistance (Horan and Shortle, 2001). Under such a policy, producers would not have to participate, although those who stand to benefit economically from a change in management may opt to do so. Such policies have been a mainstay of USDA conservation policies for decades. However, education and technical assistance have been shown to have limited effects on managing nonpoint pollution when not paired with other policy instruments (Ribaudo and Horan, 1999).

Regulations on inputs, management practices, or estimated emissions provide more direct stimulus. For a regulation on inputs or management practices (design-based policy) to be efficient, all polluting inputs used by farmers would have to be controlled (Ribaudo et al., 1999). Such regulations tend to be inflexible, and to be efficient the regulator would need to know practice costs as well as their impacts on pollutant delivery. A regulation on estimated emissions (performance-based policy) provides more flexibility to farmers, but requires models that farmers can use to estimate how their management choices affect emissions.

Farmer decisions can also be influenced through economic incentives like taxes on estimated emissions or polluting inputs, subsidies for management practices that improve water quality, and markets for trading pollution allowances. For measured or estimated delivered emissions, only a single tax (on emissions) or subsidy rate (on abatement) for each pollutant is needed to achieve an efficient allocation of abatement. For inputs, however, a system of taxes for polluting inputs and incentives for abating practices needs to be applied for the policy to be efficient, and the rates would vary for each farm to account for differences in delivery, making such an approach far more complex (Ribaudo et al., 1999). In general, incentives are more flexible than standards, as farmers choose the best way to manage their operations given the incentives.

Comparing Policy Instruments

Different policy instruments address the "Who, What, and How" questions in different ways, with implications for economic performance. Performance-based policies targeting pollution emissions are generally more efficient (Ribaudo et al., 1999), designed as they are to achieve a specific environmental performance measure rather than to simply encourage the adoption of certain management practices. The choice of how to meet the performance measure is left entirely to the farmer, who has better information than resource managers on how changes in practices affect costs. Performance-based policies include emission limits, pollution taxes, and payments for improved water quality. Properly designed, such policies automatically allocate pollution control to those farms and fields that can provide the most reductions at least cost (the "Who" and "What"). They also promote the greatest flexibility in achieving policy goals, which leads to the most efficient policies (Ribaudo et al., 1999).

Because nonpoint-source emissions cannot be observed or are too costly to measure, performance-based policies have rarely been applied to nonpoint-source pollution. Instead, nonpoint-source policies tend to be design-based, pegged to a producer's variable input use and production practices.

Policies based on expected emissions, estimated with models, provide some of the flexibility offered by performance-based policies. Point/nonpoint trading markets fall in this category, as models are used to estimate the performance of practices installed for the purpose of creating credits. The cost of developing models may be high, but once developed, the cost of applying them should be low.

Design-based regulations can be cost-effective, but this requires precise targeting of specific practices to each field. Efficiency requires that the regulator have information on the abatement costs of each farm, which is privately held and may be difficult to obtain. Efficiency is reduced the greater the heterogeneity of land, the more uniformly instruments are applied across regions, the fewer polluting inputs that are controlled, and the greater the uncertainty of the resource management agency about the performance of best management systems (BMSs).[2] Standards also limit flexibility, although they are relatively easy to implement. Regulations applied to agriculture generally take this form. Common examples include requirements for nutrient management plans, bans on applying manure to frozen fields, required use of vegetative filter strips, and limits on chemical applications in aquifer recharge zones.

Design-based economic incentives such as cost-sharing promote the voluntary adoption of desirable practices. Historically, financial assistance (cost sharing) has been the primary policy instrument for inducing farmers to adopt management practices that reduce nonpoint-source pollution. A successful and efficient financial assistance approach would require practice- and region-specific payment rates that are high enough to entice farmers to participate and to adopt the most cost-effective practices. Payments may have to continue for as long as practices are maintained (current conservation programs offer only short-term contracts). Research has found that nonstructural management practices, such as nutrient management, are more likely to be discontinued after cost-share funding ends than structural practices (Jackson-Smith et al., 2010), and that farmers often find nutrient management difficult to implement (Genskow, 2012; Osmond et al., 2012). Paying farmers to adopt practices they would not ordinarily adopt on their own could be very expensive given the improvements still needed to attain water quality goals in the Bay, and raises questions about the fairness of paying farmers for pollution abatement while regulated point sources must bear the cost of abatement. Such an approach may also be at odds with the large budgetary imbalances currently affecting Federal and State finances.

Evaluation of Policy Instruments for Meeting TMDL Goals

Our assessment of policy instruments for achieving the TMDL goals on cropland in the Chesapeake Bay watershed includes performance-based regulations (emission limits), performance-based incentives (emission taxes), and design-based standards (regulations on practices or inputs). These policies include those currently included in State Watershed Implementation Plans (WIPs) as well as some that are not being considered. The strength of our analysis is in estimating the difference in costs *between* policy scenarios, rather than in our assessment of the absolute cost of any one scenario. These differences enable us to identify which policy design features are more cost-effective. While our analysis does not mimic the State Watershed Implementation Plans, our estimates do provide a measuring stick for the costs of reducing agricultural nonpoint-source pollution by an amount broadly consistent with what is being asked of agriculture to meet Bay water quality goals. The estimates reflect a stylized assessment in which the watershed is treated as one political jurisdiction and provide policymakers with information about the extent to which alternative policy approaches might deviate from the least-cost solution.

Data

The NRCS Conservation Effects Assessment Project (CEAP) for the Chesapeake Bay watershed provides the data for our crop sector analysis (see box, "CEAP Data"). The data consist of 756 field-level survey observations from 2003-2006 containing information on baseline crop acreages, input use, conservation practices, estimated edge-of-field nutrient and sediment losses, and estimated delivery of those materials to the tidal waters of the Bay. In addition, the data contain estimated nutrient and sediment delivery for three "enhanced" management practices—cover crops, nutrient management, and water erosion controls—that could be added to baseline practices. The 756 sample points are treated as representative farms that capture regional variation in resource endowments and base technology use. We use these representative farms to evaluate a suite of performance- and design-based policy options for meeting the TMDL (see box, "Model Management Practice Choices"). Baseline observations without any of these practices could receive all the practice combinations. Baseline observations already receiving the full suite of practices would not receive any additional practices.

Over 57 percent of cropland in the Chesapeake Bay watershed was not under any of the three management systems in 2006 (table 2). About 20 percent had only water erosion controls, and 12 percent had only nutrient management. Very few cropland acres had more than one system in place prior to the TMDL. (A follow-up CEAP survey was conducted in 2011 to track changes in management between 2006 and 2011, see box, "2011 CEAP Update.")

Up to 8 different sets of conservation systems are possible for each of the 756 sampling points, depending on baseline practices:

- No enhanced conservation system
- Cover crops
- Nutrient management practices
- Water erosion control
- Water erosion control plus cover crops
- Water erosion control plus nutrient management
- Nutrient management plus cover crops
- Nutrient management plus cover crops plus water erosion control.

CEAP Data

The Conservation Effects Assessment Project (CEAP) was initiated in 2002 by USDA's Natural Resources Conservation Service (NRCS), Agricultural Research Service, and the National Institute of Food and

Agriculture to evaluate the benefits from resource conservation on agricultural lands. A series of large watershed-scale studies was conducted as part of CEAP to focus on the water quality benefits from conservation systems. The focus of the CEAP Chesapeake Bay study is on the 4.4 million acres of cultivated cropland in the watershed.

A survey of farmers obtained information on the extent of conservation practice use in the watershed over 2003-2006. The National Resources Inventory (NRI)—a statistical survey of conditions and trends in U.S. soil, water, and related resources on private lands conducted by NRCS since the 1970s—provided the statistical framework. The CEAP-NRI survey collected detailed production and conservation management data for crop fields in the NRI over crop years 2001-2006. The statistical properties of the observations capture the distribution of practices on all agricultural lands within each four-digit hydrologic unit code (HUC) in the Bay watershed. We restricted our analysis to major field crops, which constituted 97.9 percent of crop acres in the watershed. We excluded fields planted to vegetables, as we have less confidence that our cost data are representative of these specialized crops. This left us with 756 observations.

The CEAP data also contain model estimates of erosion and nutrient emissions from fields under both the baseline and alternative management systems. Field-level effects of conservation practices were assessed using the Agricultural Policy Environmental Extender (APEX) (Williams et al., 2008; Gassman et al., 2009). APEX is a field-scale physical process model that simulates day-to-day farming activities, wind and water erosion, loss or gain of soil organic carbon, and edge-of-field losses of soil, nutrients, and pesticides. APEX also estimates crop yields for a given set management choices. The impact of field-level emissions on loadings of nitrogen, phosphorus, and sediment within the watershed and to the tidal waters of the Bay was simulated with the Hydrologic Unit Model for the United States (HUMUS) (Arnold et al., 1999; Srinivasan et al., 1998). Within HUMUS, the Soil and Water Assessment Tool is used to route instream loads from one watershed to another (Gassman et al., 2007; Arnold and Fohrer, 2005).

Weather is a predominant factor in determining the emissions of nutrients and sediment from cropland. To capture the effects of weather, each scenario was simulated using 47 years (1960-2006) of daily weather data. The estimates used in our analysis are the means of these simulations.

The field-level effects of current conditions and the alternative management scenarios on edge-offield nitrogen, phosphorus, and sediment losses (and delivery to the tidal waters of the Bay) are assessed for each observation by NRCS using a system of models (see box, "CEAP Data")[3]. The modeling accounts for any interactions from adopting multiple practices.

Model Management Practice Choices

Cover crops. Cover crops are planted when the principal crops are not growing to provide soil surface cover and reduce erosion, and to withdraw excess nutrients remaining in the soil, preventing them from leaching or running off to surface water. In our model:

- For sample points in the baseline without a cover crop in the rotation, rye was planted as the cover crop.
- The cover crop is planted the day after harvesting the main crop or the day after the last major fall tillage practice.
- The cover crop is not harvested for sale.

Nutrient management. Nutrient management is defined in terms of the appropriate rate, timing, and method of application for all crops in the rotation. In our model:

- All commercial fertilizer is applied 14 days prior to planting, except for acres susceptible to leaching loss.
- For acres susceptible to leaching, nitrogen was applied in split applications.
- Manure applications during winter months are moved to the spring.
- All fertilizer and manure are incorporated or injected.
- All nitrogen application rates for all crops except cotton and small grains are limited to 1.2 times the crop removal rate. For small grains, nitrogen applications are limited to 1.5 times the crop removal rate. For cotton, nitrogen applications are limited to 50 pounds per bale.
- Phosphorus application rates are adjusted to be equal to 1.1 times the amount removed in the crop at harvest.

Water erosion control. Water erosion controls reduce sediment loss by reducing field slopes, reducing the length of time a field is without plant cover, or trapping sediment in runoff. In our model:

- Terraces are added to all sample point with slopes greater than 6 percent, and to those with slopes greater than 4 percent and a high potential for excessive runoff (hydrologic soil groups C or D).
- Contouring or strip-cropping is added to all other fields with slopes greater than 2 percent that do not already have those practices.
- Fields adjacent to water receive a riparian buffer.
- Fields not adjacent to water receive a filter strip.
- No changes are made to tillage.

Table 2. Crop acres under different conservation management systems in the baseline

Management system	Acres	Percent
None	2,408,685	57.5
Nutrient management (NM)	521,018	12.4
Cover crops (CC)	119,260	2.8
Water erosion control (EC)	855,582	20.4
NM-CC	11,019	0.3
NM-EC	225,792	5.4
CC-EC	18,972	0.4
NM-CC-EC	28,583	0.7

Source: USDA, Natural Resources Conservation Service, 2011.

2011 CEAP Update

NRCS conducted a new CEAP survey in 2011 to evaluate progress made by Bay farmers between 2006 and 2011 (USDA, NRCS, 2013). The new survey captured changes made by farmers through the various Federal and State programs providing technical and financial assistance, as well as changes farmers made on their own. The survey found increased use of some conservation measures, particularly cover crops and erosion controls, in the watershed. The survey also found little change in the use of nutrient management practices. The 2011 data were not publicly available as of the publication date (or "in time" for our analysis). Had we based our analysis on the 2011 data, our estimates of total costs would probably be lower, but our conclusions about the relative cost effectiveness of the different policy approaches should not have changed.

The model simulations provide estimates without any of the uncertainty characteristics of nonpointsource pollution that are associated with variability in weather and practice performance. Since we are using the model results to compare policy instruments for reducing agricultural nonpoint-source pollution, this shortcoming is not seen as a major issue and policy rankings should not be affected. Uncertainty would be important if we were trying to determine the level of management necessary to meet a water quality standard with a degree of certainty or to compare the cost-effectiveness of abatement from sources with very different uncertainty characteristics. We assume that policymakers are concerned only with long-term average outcomes. In our modeling, we assume that farms do not enter or exit the sector in response to the introduction of TMDL-related policies. We also do not consider taking cropland out of production (shifting to pasture or forest) as an option for reducing pollution losses, other than converting cropland to filter strips as part of water erosion controls. The CEAP data do not include this practice. We also assume that policies implemented in the Bay watershed do not influence market prices, so there would be no incentive for farms not affected by an abatement policy to alter their production (no "leakage").

Cost of Best Management Systems

We assume that farmer management choices are the result of profit maximizing behavior. The cost to a farmer of implementing a new conservation practice is the expected reduction in net farm returns. These include annualized capital costs (equipment and structures), changes in annual input costs (labor, fuel, chemicals, seed), and changes to gross revenue due to changes in crop yields (including land taken out of production). Capital and annual costs were obtained from several sources, including State Watershed Implementation Plans and contract data from USDA's Environmental Quality Incentives Program. Changes in crop yields for the different management scenarios were obtained directly from the CEAP modeling. Policy transaction costs (administration, monitoring, verification, and enforcement) are not included in the analysis. We also could not include manure hauling costs incurred by confined animal feeding operations with inadequate cropland for agronomic application of the manure nutrients.

We first examine whether the management baseline represented by the CEAP survey data was economically "optimal." If we assume that farmers have perfect information and seek to maximize expected profits, we expect that observed practices are those that maximize net returns. However, we found that a small portion of our sample could increase net returns by adopting some of the improved management systems, implying that farmers are not maximizing net returns, a result found by others using profit-maximizing models that assume perfect information (Diaz-Zorita et al., 2004; Pendell et al., 2006; Veith et al., 2004). In an analysis such as ours, this could lead to a policy intervention having a negative cost.

Two factors may be at work here. First, the actual cost of adopting practices for each observation is unknown; only regional averages are available. Second, a variety of factors that operators treat as costs may not be captured by a simple accounting of implementation costs and revenues. These include increased uncertainty about expected returns (risk), loss aversion (decisions are influenced more by a potential loss than an equal-sized potential gain), and behavioral inertia (Gillenwater, 2012). Consider the split application of nutrients, which is a practice for improving nutrient use efficiency. If the field is too wet when the second application needs to be applied, then yields will suffer. A farmer may be reluctant to adopt this practice even though, on average, it offers an increase in net returns.

The data did not allow us to explore the cost and risk characteristics of each observation. Instead, we estimated what the implicit cost of the enhanced management systems would have to be for the observed management practice to be the profit-maximizing choice. For each modeled scenario, we determined the largest increase in net returns from among all the observations. We then added this value as an implicit cost to adopting the conservation measures to each observation. For example, if we require each sample farm not currently using nutrient management to adopt the practice, and a single farm shows a profit by doing so, we would add that profit as an implicit "cost of risk" to each sample farm in the database. This ensures that the baseline scenario represents those practices that give farmers in the watershed maximum net returns in the absence of any policy, while preserving the rankings of observations in terms of management system implementation costs. Changes in management would therefore only be in response to the incentives offered by the policy, and not to any other economic factor. While this approach cannot be used to estimate the "true" cost of a policy, we believe it can be used to compare policies. We refer to this addition to cost as the "risk" component throughout the rest of the report.

Performance-Based Instruments

Performance-based policies use measures of pollutant delivered to receiving waters as the basis for policy. We evaluated performance-based scenarios using economic models developed to estimate the least-cost distribution of best management system (BMS) adoption across the watershed for achieving the pollution abatement goals (see Appendix 1 – Modeling Scenarios for Improving Cropland Nutrient Management).

Bay-Level Constraint (Optimal)

This scenario represents the least-cost distribution of BMSs for achieving the nitrogen, phosphorus, and sediment abatement goals of the TMDL at the tidal waters of the Bay, given the baseline practices as the starting point. The optimization model selects the mix of conservation systems across the watershed from among those included in the CEAP data to achieve the TMDL at least cost. With complete information about pollutant delivery to tidal waters from each field and practice costs, a regulatory agency could establish a delivered pollutant limit for each field. Farmers would then determine which practices would meet the limit at least cost to them. This "optimal" scenario represents the most efficient allocation of abatement across the watershed and is useful as a comparison for the other policy instruments in terms of cost and the regional distribution of abatement.

Hydrologic Unit Code (HUC)-Level Constraint

Issues such as "fair share" and the simplification of policies may lead decisionmakers to limit their policy choice set to a manageable number of alternatives and ad hoc decision rules (Schwartz, 2010). Fair share implies a distribution of abatement across political jurisdictions such that one jurisdiction is not perceived as bearing the brunt of the costs, which is important in a watershed such as the Chesapeake Bay. This scenario assumes that political realities require all political jurisdictions (for simplicity, defined here as a 4-digit HUC, fig. 1) to have identical abatement goals, in terms of percentage reductions in nitrogen, phosphorus, and sediment. We model this scenario by constraining each of the four HUCs to achieve the same percentage reductions in delivered nitrogen, phosphorus, and sediment, while meeting the overall TMDL abatement goal for the tidal waters.

Field-Level Constraint

This scenario is the least-cost combination of BMSs necessary for each field to reduce delivered nitrogen, phosphorus, and sediment by the TMDL reduction goals. Farmers choose how to meet the requirements. This scenario treats all fields equally, regardless of location in the watershed, current emissions, or baseline management. Treating all fields the same may be seen as more "fair" than a policy that targets only a subset of fields, as resource managers may reject sharp differences in the allocation of abatement across comparable nonpoint sources (Schwartz, 2010).

HUC- and field-level requirements may be less efficient than Bay-level constraints in achieving TMDL goals at the tidal waters, but they may enable more widespread water quality improvements in local rivers and streams. The water quality benefits to local users may engender more political support from regions that do not benefit directly from improved water quality in the Bay. We do not assess location-specific economic benefits in this analysis.

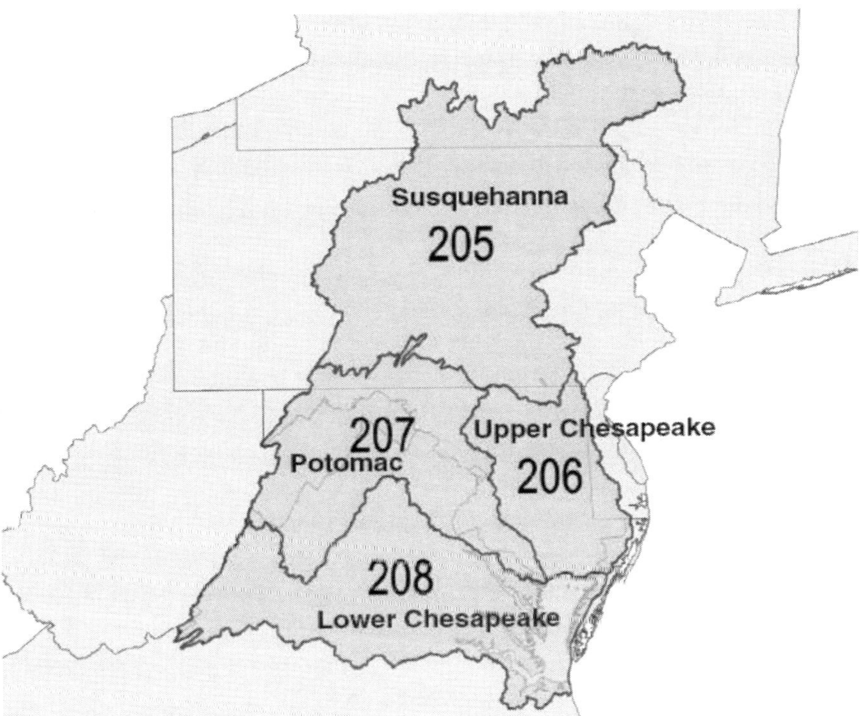

Source: USDA, Economic Research Service.

Figure 1. Chesapeake Bay Watershed and major sub-basins (HUC 4).

Design-Based Instruments

Performance-based policies are difficult to implement for nonpoint-source pollution because pollutant discharge cannot easily be measured and regulators lack the information necessary to set optimal performance goals. An alternative is to focus on inputs and management practices (design-based approaches), which are more easily observed by both resource management agencies and farmers. Practices can be required through regulation or encouraged through financial incentives. An important difference in the evaluation of these scenarios is that they are not constrained to meet the TMDL goals.

An essential design feature of these policies is identifying which fields are subject to regulation or incentives (the “Who” question). If pollution control programs could target priority fields (based on geography or field characteristics) and the farmers of those fields adopt the most efficient pollutant-reducing practices, then the total costs of abatement would be minimized (performance-based practices do this automatically through their incentive structure). The key is to find targeting criteria that are the best predictors of both water quality and low unit abatement costs, but are also easy to implement (low administrative costs). We selected targeting criteria from among the variables included in the NRCS CEAP data set that we believe could be used in such a policy.

Scenarios are evaluated based on the likely attainment of basinwide TMDL goals, total costs to agriculture, and unit abatement costs (a measure of efficiency). The costs of the design-based scenarios are estimated by selecting the observations/fields meeting the targeting conditions (described in the scenarios below), adding required management practices to the baseline practices, then adding up the total BMS implementation costs. Exceeding TMDL goals is not rewarded; as long as the goals are met, the lowest cost policy is the one preferred.

Who bears the cost of design-based instruments depends on whether practices are required through regulation (farmers and consumers bear the cost) or whether financial incentives are used to compensate farmers for lost profits (taxpayers bear the cost). We assume that a farmer will be willing to adopt a practice if total implementation costs are covered, so that the costs of regulation and financial incentives would be the same.

Mandatory Best Management Systems (BMS) on All Acres (E3)

This scenario simply requires all crop acres to implement the full suite of best management systems for improving water quality: nutrient management (NM), water erosion controls (EC), and cover crops (CC) (Everyone does Everything Everywhere, or E3). This approach does not require resource management agencies to determine baseline pollutant losses or rank the effectiveness of practices across cropping systems and resource characteristics, as targeting is not a part of this policy. The expectation is that this approach would be the most inefficient (Wainger, 2012).

At Least One BMS on All Acres

This scaled-down approach (relative to the E3 scenario) requires that all cropland acres be covered by at least one best management system—nutrient management, water erosion controls, or cover crops. Only fields that have not implemented any of these systems have to adopt one, at the discretion of the farmer. We assume farmers adopt the least-cost system. This scenario recognizes good stewards who have already adopted water quality-protecting systems and focuses on those producers slow to adopt them. The scenario also demonstrates how farmers might respond to a voluntary program that *fully* offsets adoption costs of management systems selected by farmers (otherwise farmers have little incentive to adopt the systems). Most conservation programs provide only partial cost-share assistance (EQIP cost shares are generally around 50 percent). However, some programs, such as Maryland's Cover Crop program, offer 100-percent cost shares.

BMSs Targeted to Vulnerable Fields

This scenario uses information on the inherent vulnerability of fields to runoff or leaching in order to target the policy. Vulnerability factors for surface runoff include soil properties such as hydrologic group, slope, and erodibility (USDA, NRCS, 2011). Factors for loss of nutrients to subsurface flow include soil hydrologic group, slope, erodibility, and coarse fragment content of the soil. About 23 percent of cropped acres in the Chesapeake Bay watershed have high runoff potential, and 42 percent have moderate runoff potential (USDA, NRCS, 2011). About 17 percent of cropped acres have high leaching potential and 46 percent have moderate leaching potential. Vulnerable acres are not distributed evenly across the watershed. About 60 percent (53 percent) of acres that are highly or moderately vulnerable for runoff (leaching) is located in the Susquehanna basin (table 3). In this scenario, if only leaching vulnerability is high, then nutrient management and cover crops are required. If only runoff vulnerability is high, then cover crops and erosion control are required. If both leaching and vulnerability are high, then all three management systems are required.

BMSs Targeted to Fields Needing Treatment

Targeting on inherently vulnerable cropland ignores the fact that the land may have adequate treatment to address pollutant loss. This scenario uses vulnerability factors and baseline management practices to identify those fields that are inadequately treated, as defined by NRCS (USDA, NRCS, 2011). NRCS estimates that 19 percent of cropped acres had a high level of need for additional conservation measures in 2006, and 61 percent had a moderate need (80 percent undertreated in total). Almost 70 percent of acres with the highest treatment needs are in the Susquehanna basin (table 3). This basin contains 56 percent of the CBW acres receiving manure and about 67 percent of the acres classified as highly vulnerable to runoff, and as a result, contributes a disproportionate share of pollutants. In this scenario, only fields defined as needing treatment receive additional water quality BMSs. Based on NRCS ratings, nutrient management (NM) and cover crops (CC) are required if there is a need to reduce nitrogen leaching or runoff. NM and erosion controls (EC) are required if there is a need to reduce phosphorus runoff. EC and CC are required if there is a need to reduce sediment loss. If two or more pollutant problems need addressing, then nutrient management, erosion controls, and cover crops are required. We allow the model to select the least-cost system rather than prescribe it. This represents how farmers might respond to a voluntary, targeted program that compensates them for adopting a management system.

Table 3. Crop acres, vulnerable acres, and acres in need of treatment, by sub-basin (HUC-4)

Cropland category	HUC 205 Susquehanna	HUC 206 Upper Chesapeake	HUC 207 Potomac	HUC 208 Lower Chesapeake	Total
All cropland	1,734,800	1,187,900	684,000	673,200	4,279,900
Vulnerability					
High runoff potential	654,792	65,358	188,640	70,535	979,325
Moderate runoff potential	998,411	229,610	397,988	165,456	1,791,465

Cropland category	HUC 205 Susquehanna	HUC 206 Upper Chesapeake	HUC 207 Potomac	HUC 208 Lower Chesapeake	Total
High leaching potential	322,417	195,497	93,583	105,938	717,435
Moderately high leaching potential	1,097,013	421,369	295,657	152,562	1,966,601
Treatment need					
Critical under-treated acres	558,625	92,499	120,517	38,356	809,997
All under-treated acres	1,464,251	862,878	590,263	490,533	3,407,924
Enhanced targeting factors					
Adjacent to water	323,992	275,733	190,378	145,746	935,849
Highly Erodible Land					
(HEL)	1,099,996	236,129	388,458	153,532	1,878,115
Cropland receiving manure	913,625	401,765	293,993	7,805	1,617,188

HUC = Hydrologic Unit Code designation for a watershed
Source: USDA, Natural Resources Conservation Service, 2011.

BMSs Targeted to Fields Adjacent to Water

Fields adjacent to water contribute more pollutants than upland fields farther away, all else equal (Baker et al., 2007; Kling, 2011). About 22 percent of cropland in the Chesapeake Bay watershed is identified as being adjacent to water (table 3). About 35 percent of this land is in the Susquehanna watershed and 29 percent in the Upper Chesapeake watershed. We explore different combinations of nutrient management, erosion controls, and cover crops to determine which would achieve TMDL goals at the lowest cost. We also allow targeted farmers to select the least-cost management system themselves.

BMSs Targeted to Fields with Highly Erodible Land

Highly erodible land (HEL) is subject to high runoff levels, which carry with it sediment and nutrients. Cropland is routinely assessed for erosion since HEL cropland is subject to USDA compliance provisions. About 44 percent of cropland in the Bay watershed was identified as HEL in 2006 (table 3). About 60 percent of this HEL was in the Susquehanna watershed. We explore different combinations of management practices to find which could meet the TMDL goals at least cost when targeted to HEL cropland. We also allow targeted farmers to select the least-cost management system.

BMSs Targeted to Fields Receiving Manure

Cropland receiving manure has consistently been found to receive excess amounts of nutrients (Ribaudo et al., 2003; Ribaudo et al., 2011). Part of this is due to the need to dispose of manure on operations that have large numbers of animals. Manure is also more expensive to apply in a conserving manner than commercial fertilizer. About 38 percent of cropland in the watershed received manure in 2006 (table 3). About 56 percent of land receiving manure was in the Susquehanna watershed. We explore different combinations of management practices to find which could meet the TMDL goals at least cost when targeted to cropland receiving manure. We also allowed targeted farmers to select the least-cost management system.

Results

Table 4 describes each policy scenario for achieving pollution abatement in the Chesapeake Bay watershed: whether there is a pollutant reduction constraint, whether farmers are "told" what to implement and what those requirements are, and whether the policy is targeted to certain kinds of fields. The scenarios were evaluated with an optimization model described in Appendix I.

The ***Bay-level policy scenario***, which minimizes the cost of simultaneously meeting the nitrogen, phosphorus, and sediment reduction goals at the tidal waters of the Bay, was the optimal solution, contingent on resource managers having perfect information on costs and the delivery of pollutants from each field. The cost of meeting TMDL abatement goals was estimated to be about $54.8 million per year (table 5). This cost estimate includes our measure uncertainty about expected returns (risk), loss aversion, and behavioral inertia which are unobserved but reflect real costs to the farmer and thus influence their practice adoption decisions. Unit abatement costs were $2.13 per delivered pound for nitrogen, $67.27 per delivered pound for phosphorus, and $647.33 per delivered ton for sediment. The pollutant reduction goals were met by adopting conservation systems on only 505,000 acres (12 percent of cropland in the watershed) (table 6). Water erosion controls and nutrient management were applied to over 90 percent of the cropland receiving additional treatment. Cover crops were applied to less than half of cropland receiving treatment. Over 56 percent of the crop acres with improved management were located in the Susquehanna sub-basin (HUC 205), which is the largest source of pollutants to the Bay. Pollution can be reduced at a lower unit cost on average in this sub-basin than in the others. Only 1.7 percent of crop acres with improved management were located in the Lower Chesapeake (HUC 208).

Table 4. Summary of policy scenarios

Policy	Constrained to meet the N, P, and S reduction goals?	Farmers told what to implement?	Targeted	Practice requirement
	Performance based			
Bay level (optimal)	Yes, at the Bay	no	no	None, farmer can choose least cost combination of practices to meet goal
HUC-4 level	Yes, at the HUC	no	no	None, farmer can choose least cost combination of practices to meet goal
Field level	Yes, at the field	no	no	None, farmer can choose least cost combination of practices to meet goal
	Practice based			
Every field has every practice (E3)	no	yes	no	Every farmer uses every practice
Every field has at least one practice	no	no	no	Fields without any practices in the baseline adopt one practice
Vulnerability	no	yes	yes	CC and NM if vulnerable to leaching CC and EC if vulnerable to runoff
Treatment need	no	yes	yes	NM and CC if need to reduce N leaching or runoff. NM and EC if need to reduce P runoff.

Table 4. (Continued)

Policy	Constrained to meet the N, P, and S reduction goals?	Farmers told what to implement?	Targeted	Practice requirement
				EC and CC if need to reduce sediment loss. NM, EC, and CC if two or more needs.
Treatment need: least cost practice	no	no	yes	None, targeted farmers adopt least cost practice of their choice
Adjacent to water and treatment need	no	yes	yes	EC and NM on cropland adjacent to water and with a moderate treatment need, AND on croplandwith a high treatment need
Adjacent to water and treatment need: least cost practice	no	no	yes	None, targeted farmers adopt least cost practice of their choice
HEL	no	yes	yes	EC and NM on HEL cropland
HEL: least cost practice	no	no	yes	None, targeted farmers adopt least cost practice of their choice
Manure and treatment need	no	yes	yes	EC and NM on cropland receiving manure with a moderate treatment need, AND on cropland with ahigh treatment need
Manure and treatment need: least cost practice	no	no	yes	None, targeted farmers adopt least cost practice of their choice

CC = cover crop
NM = nutrient management
EC = water erosion control
HEL=Highly Erodible Land

Table 5. Cost, pollutant reduction, and unit cost of policy instruments by scenario

Policy	Cost ($)	Nitrogen reduction (%)	Phosphorus reduction (%)	Sediment reduction (%)	Unit cost N ($/lb.)	Unit cost P ($/lb.)	Unit cost Sediment ($/ton)
Performance-based							
Bay level (optimal)	54,775,571	25	24	20	2.13	67.27	647.33
HUC-4 level	80,388,269	25	24	21	3.12	97.18	901.01
Field level	801,783,748	58	58	72	13.43	408.83	2,636.96
Design-based							
Every practice on all fields (E3)	1,264,018,051	75	74	85	16.39	501.13	3,497.87
Every field has at least one practice	399,922,550	24*	30	36	16.41	397.00	2,623.07
Vulnerability	420,960,863	28	30	34	14.69	419.16	2,953.07
Treatment need	949,140,202	64	63	63	14.40	447.01	3,562.09
Treatment need: least cost practice	531,305,469	28	40	54	18.21	393.70	2,324.78
Enhanced targeting							
Adjacent to water and treatment need: EC and NM	337,299,175	30	31	25	10.95	318.72	3,252.61
Adjacent to water and treatment need: least cost practice	210,604,209	17*	21*	23	11.64	295.56	2,144.47
HEL. EC and NM	479,113,517	34	39	39	13.53	361.37	2,873.07
HEL: least cost practice	290,933,506	21*	26	37	13.32	328.62	1,863.52

Table 5. (Continued)

Policy	Cost ($)	Nitrogen reduction (%)	Phosphorus reduction (%)	Sediment reduction (%)	Unit cost N ($/lb.)	Unit cost P ($/lb.)	Unit cost Sediment ($/ton)
Manure and treatment need: EC and NM	445,610,300	40	32	26	10.91	407.52	4,082.59
Manure and treatment need: least cost practice	263,715,463	20*	28	25	12.27	284.62	2,473.32

*Scenario did not meet Total Maximum Daily Load goals.

CC = cover crop.

NM = nutrient management.

EC = water erosion control.

HEL = Highly Erodible Land.

Source: USDA, ERS analysis of Natural Resources Conservation Service Conservation Effects Assessment Project data.

One way that this optimal abatement could be achieved would be for the regulatory agency to assign a ***delivered discharge limit*** for each field. A discharge limit allows each farm to meet the standard at least cost, but allocative efficiency requires that each field have a unique limit that accounts for practice costs. Such an approach would be very costly to implement, even if the data were available.

An alternative is to implement a ***tax on pollutants*** discharged to the tidal waters. The estimated taxes on delivered pollutants to simultaneously achieve the necessary reductions are $2.36 per pound N, $39.98 per pound P, and $200.08 per ton sediment.[4] We assume that any tax revenue would be rebated back to farmers through a system of awards for adopting conservation systems. To illustrate the incentive provided by an emissions tax, we look at farms that had none of the best management systems in place prior to the TMDL. The annual tax burden on these farms would be about $130 million. By adopting all three management systems, their tax burden would be reduced by $102 million. As long as the implementation cost of a management system is less than the proposed pollutant tax, the farmer would implement the system.

Table 6. Practice adoption by scenario

Policy	No change	CC	NM_EC_CC	EC_CC	NM_EC	EC	NM_CC	NM	Total changed
	Acres								
Performance-based									
Bay level (optimal)	3,683,919	37,689	51,323	121,418	204,522	41,150	31,025	17,862	504,990
HUC-4 level	3,573,546	26,109	112,144	136,948	175,972	112,911	26,672	24,606	615,363
Field level	7,729	652,297	304,466	2,098,949	403,884	585,273	111,704	24,606	4,181,180
Design-based									
E3	28,583	225,792	2,408,685	521,018	119,260	11,019	855,582	18,972	4,160,326
Every field has at least one practice	1,780,224	692,917	0	487,062	21,204	1,156,057	8,976	42,468	2,408,685
Vulnerability	2,641,684	172,186	102,196	746,630	0	36,183	475,229	14,802	1,547,225
Treatment need	1,039,332	42,004	954,194	112,905	370,050	11,500	1,452,615	206,310	3,149,577
Treatment need: least cost practice	865,509	766,387	0	754,333	21,204	1,730,031	8,976	42,468	3,323,400
Enhanced targeting									
Adjacent to water and treatment need: EC and NM	2,829,993	0	0	0	1,001,205	42,054	0	315,658	1,358,917

Table 6. (Continued)

Policy	No change	CC	NM_EC_CC	EC_CC	NM_EC	EC	NM_CC	NM	Total changed
Adjacent to water and treatment need: least cost practice	2,825,204	445,609	0	220,608	21,204	649,446	8,976	17,862	1,363,705
HEL:EC and NM	2,433,306	0	0	0	1,167,012	182,713	0	405,878	1,755,603
HEL: least cost practice	2,359,348	701,622	0	206,253	21,204	850,588	25,598	54,297	1,829,561
Manure and treat-ment need: EC and NM	2,495,461	0	0	0	1,285,489	73,197	0	334,762	1,693,448
Manureand treat-ment need: least cost practice	2,482,736	535,361	0	356,397	21,204	766,374	8,976	17,862	1,706,173

CC = cover crop.

NM = nutrient management. EC = water erosion control.

HEL = Highly Erodible Land.

Source: Analysis of USDA, Natural Resources Conservation Service Conservation Effects Assessment Project data.

The optimal solution concentrated pollutant reductions and associated costs in the Susquehanna watershed. This might seem unfair, particularly since the Susquehanna watershed does not border the Bay and residents would not directly benefit from the Bay's improvement. A more "equitable" requirement could be that each basin reduces delivered nitrogen, phosphorus, and sediment by the same percentages. However, shifting delivered abatement to parts of the watershed characterized by higher unit costs increases the overall costs of meeting the TMDL by almost 47 percent (HUC-4 level in table 5). Unit abatement costs increase to \$3.12 per pound for nitrogen, \$97.18 per pound for phosphorus, and \$901.01 per ton for sediment. About 22 percent more crop acreage is treated than under the Optimal scenario. The share of treated acres coming from the Susquehanna sub-basin decreases to 45 percent, while the share coming from the Lower Chesapeake increases from 1.7 percent to 19.5 percent (table 7).

Table 7. Distribution of treated acres by sub-basin (4-digit Hydrologic Unit Code (HUC))

Policy	HUC 205 Susque-hanna	HUC 206 Upper Chesapeake	HUC 207 Potomac	HUC 208 Lower Chesapeake
	Percent of cropland			
Performance-based				
Bay level (optimal)	56.6	21.7	20.0	1.7
HUC-4 level	45.3	22.8	12.4	19.5
Field level	40.4	27.3	16.3	16.0
Design-based				
E3	42.3	25.7	15.3	16.7
Every field has at least one practice	42.9	24.6	16.7	15.9
Vulnerability	56.9	15.2	16.5	11.4
Treatment need	43.6	24.1	17.2	15.2
Treatment need: least cost practice	42.7	24.8	17.7	14.8
Enhanced targeting				
Adjacent to water and treatment need: EC and NM	54.2	16.4	19.4	10.0

Table 7. (Continued)

Policy	HUC 205 Susque-hanna	HUC 206 Upper Chesapeake	HUC 207 Potomac	HUC 208 Lower Chesapeake
Adjacent to water and treatment need: least cost practice	54.3	16.3	19.4	10.0
HEL:EC and NM	58.9	11.1	21.5	8.4
HEL: least cost practice	58.3	12.2	21.1	8.4
Manure and treatment need: EC and NM	57.1	21.2	19.2	2.6
Manure and treatment need: least cost practice	57.1	21.0	19.3	2.6

CC = cover crop.

NM = nutrient management. EC = water erosion control. HEL = Highly Erodible Land.

Source: Analysis of USDA, Natural Resources Conservation Service Conservation Effects Assessment Project data.

Requiring that the watershed pollutant abatement goals be met at the field level proved far more costly (table 5). Total control costs were more than 14 times greater than for the Optimal scenario. Additional conservation practices were adopted on nearly all cropland (table 6). As a result, watershed-level abatement goals were significantly exceeded (58-percent reduction for N and P, and 72-percent reduction for sediment). This is partly due to the "all or nothing" nature of the conservation systems in our analysis; the conservation systems could not be fine-tuned to deliver specific abatement levels. The distribution of cropland acres receiving additional treatment closely matches that of the distribution of cropland in the watershed, rather than aligning with the sources of pollutants delivered to the Bay.

Design-Based Policies

Requiring that all crop acres be treated with the ***full suite of management practices (E3)*** is a simple approach to implement. It is also the most expensive (23 times more expensive than the Optimal scenario), as it makes no attempt to make use of information on vulnerability to pollutant loss or current levels of management (table 5). Not surprisingly, the pollution abatement goals are exceeded by wide margins: 75-percent reduction for N, 74 percent for P, and

85 percent for sediment. These results indicate that ample opportunities exist to reduce costs through targeting.

Requiring that all crop acres be treated with ***at least one water quality BMS*** is a lower cost alternative to the E3 approach. It is seven times more expensive than the Optimal scenario but about 30 percent of the cost of the E3 scenario, as only those fields not under any BMS would be treated (table 5). A drawback of this approach is that the abatement goal for nitrogen is not met. Farmers choose practices that are most beneficial (less costly) to them and not necessarily consistent with watershed goals.

A more efficient approach may be to target on factors believed to be closely related to pollution creation and delivery (Ribaudo et al., 1999). Requiring selected ***management systems on cropland that NRCS has identified as vulnerable*** to pollutant losses more than meets the TMDL abatement goals (N reduced 28 percent, P reduced 30 percent, and sediment reduced 34 percent) (table 3-4). About 1.6 million acres of cropland receive additional conservation practices in this scenario (table 3-5). Total costs are reduced by almost 67 percent compared to the E3 scenario. However, costs are still almost eight times greater than for the Optimal (performance-based, Bay-level) scenario. The distribution of crop acres receiving additional treatment is reasonably close to that of the Optimal scenario, although the Lower Chesapeake sees a significantly higher percentage of cropland treated (11.4 percent) than in the Optimal scenario (1.7 percent) (table 7).

Targeting only vulnerable acres requires some land that has adequate treatment in the baseline to produce additional abatement. Additional abatement can only be achieved at a high cost, reducing the overall efficiency of the policy. Costs are likely to be reduced by targeting on ***NRCS-defined treatment needs***, which takes into account both vulnerability and baseline practices. Targeting only cropland with a high treatment need (not shown) did not meet TMDL abatement goals for any of the pollutants. Expanding coverage to include cropland with high or moderate treatment needs greatly exceeded the abatement goals: 64-percent reduction for N, 63 percent for P, and 63 percent for sediment (table 5). Over 3 million acres of cropland undergo improved management, about 86 percent more cropland than under the vulnerability scenario (table 6). This increase in treated acres is the reason why total costs are much higher than targeting based on vulnerability. Unit reduction costs, which are a measure of cost-effectiveness, are not definitively better than those under the Vulnerability scenario (better for N, worse for P and sediment). But the fact that overall pollutant reductions were more than doubled without a large increase in average unit abatement cost indicates there are some efficiency gains by targeting on treatment need.

Requiring cropland identified as having moderate to high ***treatment needs to adopt at least one additional water quality BMS of the farmer's choice*** offers a lower cost alternative to prescribing specific practices for the targeted acres. Abatement goals are met and total control costs decline by 44 percent compared to the Treatment Need scenario. Unit abatement costs are higher for N but lower for P and sediment. However, targeting on Vulnerability still meets the overall abatement goal at a lower cost. The Vulnerability scenario better matches the distribution of abatement across HUCs of the Optimal scenario than does the Treatment Need scenario, meaning that conservation measures are more likely to be cost-effective at reducing pollution (table 7).

Including additional information in the targeting criteria could further improve the efficiency of targeted policies. Targeting only ***cropland adjacent to a stream*** did not meet the unit abatement goals, even when all three management systems were required (not shown). Additional targeting criteria are needed to reach enough cropland to achieve the watershed abatement goals. A policy that requires nutrient management and soil and water controls on ***cropland that is adjacent to a stream AND has at least a moderate treatment need, or has a high treatment need anywhere in the watershed***, was found to exceed the abatement goals at a cost that is much lower than targeting based only on treatment needs or even vulnerability. Nitrogen, phosphorus, and sediment reductions are 30, 31, and 25 percent, respectively. About 1.36 million acres (32 percent of cropland in the watershed) are targeted, which is less than half the acres treated under the simple Treatment Need scenario. Unit abatement costs for N and P are significantly lower than under any of the design-based scenarios (table 5). The distribution of cropland receiving additional treatment is very close to that of the Optimal scenario, which contributes to higher cost-effectiveness (table 7). Trying to improve the efficiency of this targeting approach by letting farmers with cropland meeting the targeting criteria select a BMS does not meet the abatement goals for N and P (table 5).

Targeting highly erodible land (HEL) or land receiving manure could also meet the TMDL goals at a lower cost than targeting only on treatment needs, but at higher costs than for targeting land adjacent to water. Allowing farmers to select a practice did not meet the abatement goal for nitrogen in either of these scenarios.

Summary

TMDL goals could be met by installing management systems on a relatively small share (about 12 percent) of cropland in the Chesapeake Bay watershed, based on our data and assumptions that regulators have the information necessary to set the appropriate performance standards or emission taxes. This finding (Optimal scenario) demonstrates the unequal contribution of nonpoint-source pollutants across fields.

Focusing on a small percentage of cropland could raise concerns about fairness and hinder agreement on a common policy across political jurisdictions. However, increasing the amount of cropland required to adopt key management systems reduces overall efficiency. Such a tradeoff between equity and efficiency may be necessary to achieve political support for policies that meet TMDL requirements.

If the information required to implement such a performance-based policy is either not available or too costly to collect, practice-based regulations or incentives can be used. The results of the Optimal scenario demonstrate the cost implications of being able to target abatement to those areas that can provide the most cost-effective control. The most cost-effective design-based policy among our scenarios is to target cropland that has moderate treatment needs and is adjacent to water plus cropland with high treatment needs anywhere in the watershed. In general, using additional information (like proximity to water) in an enhanced targeting scheme achieved lower unit abatement costs than any of the other design-based scenarios that focused only on vulnerability or treatment needs. One reason is that the distribution of targeted cropland across HUCs is similar to the distribution of treated acres in the Optimal scenario (table 3-6). There may be additional targeting criteria that could lower the overall costs of policies based on highly erodible land (HEL) or use of manure. The least efficient policy (E3) was to require that all cropland use the full suite of conservation systems.

The benefit of performance-based policies are also apparent in the per-acre costs of implementing required management systems. Since performance-based policies select those fields that can produce the most abatement at least cost, fewer management systems need to be adopted. The average per-acre treatment cost for cropland receiving treatment in the Optimal scenario is \$108. For the design-based scenarios, average per-acre treatment costs range from \$160 to \$304. Because cropland could not be targeted to the same land in the Optimal scenario through relatively simple targeting criteria, more treatment systems were needed per acre in order to meet the abatement goals.

Requiring the use of improved management systems through regulation has a level of certainty in practice adoption that cannot be attained through purely voluntary approaches. If farmers are willing to adopt practices for a payment that just covered adoption costs (including reduced yields), the costs of regulation and financial incentives would be similar. However, if farmers are reluctant to change practices, incentive offers may have to be much higher than implementation costs to induce voluntary change. If so, the voluntary approach may end up costing much more than a regulation to achieve TMDL goals, assuming widespread compliance and low enforcement costs.

When farmers were allowed to select the management systems to adopt, with or without compensation, they tended not to adopt systems that would achieve all three water quality goals. Cover crops and water erosion controls tended to be preferred over nutrient management. Farmer reluctance to voluntarily adopt nutrient management has been noted in other studies (Genskow, 2012; Osmond et al., 2012). TMDL goals were achieved when allowing farmers to choose which practice to implement only when all cropland with moderate and high treatment needs was targeted. This is due to the large amount of cropland covered, reflected in the relatively high total cost ($531 million) of this strategy.

If required management practices increase operating costs to the point where a profit cannot be made, then a farm would likely go out of business and the land would be incorporated into a more efficient farm or shift to some other use. We did not evaluate the potential for the TMDL to affect land use, primarily because of the limited management system changes we included in our analysis, and our inability to accurately estimate farm-level financial impacts, including risk. In reality, farmers have many more management choices available to them than we modeled in this analysis, such as conservation tillage and new crop rotations, which would likely lower the overall economic impacts of the TMDL.

WATER QUALITY TRADING BASELINES AND THE COSTS OF MEETING TMDL GOALS

The previous section demonstrated the benefits of a performance-based approach that provides incentives for farmers of fields that can provide the most abatement at least cost to adopt appropriate management systems. One such policy is point/nonpoint water quality trading. State Watershed

Implementation Plans for meeting the Bay TMDL all include the development of point/nonpoint water quality trading programs. TMDL wasteload allocations for point sources such as Publicly Owned Treatment Works (POTWs) may require costly upgrades to meet their new emission limits. To help reduce the costs of meeting these limits, the U.S. EPA is encouraging the development of water quality trading programs that allow regulated sources such as POTWs to meet their permit requirements by purchasing offsets or credits from unregulated nonpoint sources such as agriculture. In 2006, U.S. EPA and USDA signed a Water Quality Credit Trading Agreement stating that pointnonpoint trading will complement existing subsidies for water quality improvement from agriculture (U.S. Department of Agriculture, 2006).

How Trading Works

Water quality trading is a type of emissions trading. Trading can in theory allow firms with different pollution abatement costs to allocate pollution abatement among themselves in the most efficient way. In a cap-and-trade program, trading is organized around the creation of a good called an emission allowance, which allows the holder to discharge one unit of pollutant. Regulated dischargers— point sources that discharge directly to water through a pipe or other conveyance—can legally discharge only as much as their holding of allowances. Emissions in excess of the allowances result in a fine or other penalties. The total number of emission allowances allocated to regulated dischargers in a watershed is set by the State regulatory agency, based on the proportional contribution of the regulated pollutant to the total emissions limit for a watershed (a Total Maximum Daily Load, or TMDL, is often the basis for a water quality trading program). Emission allowances are initially allocated to firms by the regulatory agency, usually in proportion to current emissions.

Assuming that the number of allowances granted to a firm is less than its baseline emissions, all firms face a choice. Firms with discharge levels greater than their initial holdings of allowances will have to either purchase more allowances or reduce emissions. Firms with marginal costs of pollution control greater than the price of an allowance will purchase allowances in the market. Firms with marginal control costs less than the price of an allowance will reduce emissions and sell their excess allowances in the market. In equilibrium, the water quality goal is met (total emissions are no more than the number of emission allowances in the market), those firms that can reduce emissions at the lowest cost provide most of the pollution control, and marginal control costs for all emissions are equal (thereby meeting the efficiency goal).

The promise of emissions trading, along with the real-world success of air emissions trading, has led U.S. EPA to encourage States to consider agriculture as a source of offsets in water quality trading programs, and a number of States are either implementing or considering water quality trading programs that allow point/nonpoint source trading. Allowing agriculture to supply credits in a market created by capping emissions from regulated point sources would (1) reduce the cost of meeting water quality goals by allocating abatement more efficiently among sources (the economic rationale for trading); and (2) provide a strong economic incentive for unregulated nonpoint sources to play a larger role in improving water quality (U.S. EPA, 2008). This is an important goal for water quality trading in the Chesapeake Bay watershed.

Regulators designing point/nonpoint trading markets must contend with uncertainty about sources and levels of emissions, the effectiveness of best management practices, the water quality impacts of emissions from different sources, and farmer willingness to participate in a market driven by regulation (on point sources) (Hoag and Hughes-Popp, 1997; King, 2005; King and Kuch, 2003; Woodward and Kaiser, 2002; Ribaudo and Gottlieb, 2011; Horan and Shortle, 2011; Shortle 2013). The failure of current trading programs to perform as hoped can largely be attributed to failures of market design and program rules to adequately address these issues, and to the high transaction costs due to the necessity of incorporating uncertainties into market design.

The Role of Trading Baselines

In a point/nonpoint trading program, pollutant reductions that would have occurred as a result of technology upgrades by regulated point sources are replaced by equivalent, long-term reductions from unregulated nonpoint sources that would not have occurred in the absence of the trading market. In order for water quality to be protected, the nonpoint-source reduction in pollutants must be identical in quality to the forgone point-source reduction and must not have occurred without the incentive provided by the market (the credit is "additional"). This is the role of the baseline. A baseline is a prediction of the amount of pollutant emissions from an activity resulting from the expected future behavior of actors absent any policy intervention (trading program), holding all other factors constant (Marshall and Weinberg, 2012; U.S. EPA, 2007). Fields not meeting the baseline criteria cannot produce credits for sale in a market until the baseline criteria are achieved. If a full credit is issued for an activity that would have occurred anyway, then it is non-additional and total pollutant loadings will be greater than if the credit had been additional (Marshall and Weinberg, 2012).

In the Chesapeake Bay watershed, Pennsylvania, Maryland, and Virginia have all established baselines that do not reflect current practices, but that require a certain level of stewardship be attained before credits can be sold. These are consistent with U.S. EPA guidance on baselines. U.S. EPA's trading policy states that where a TMDL is in place, the Load Allocation or other appropriate baseline should serve as the threshold for nonpoint sources to generate credits (U.S. EPA, 2007). In a sense, the baseline becomes a representation of the TMDL Load Allocation implemented at the field scale, akin to the Waste Load Allocation for point sources being incorporated into the individual NPDES permits. Establishing a stewardship-based baseline essentially requires a farm to contribute to the TMDL's Load Allocation before being able to sell credits. The ability to sell credits is expected to be a sufficient incentive for farmers to provide this extra abatement. Farmers already meeting the baseline, who could be considered good stewards for previously adopting conservation practices, would be able to enter the trading program without first contributing to Load Allocation. Farmers who are not using best management systems (from a resource conservation standpoint) would likely have to adopt some conservation measures to meet baseline requirements. The more stringent the baseline (i.e., the smaller the pollutant losses allowed), the greater the percentage of cropland needing improved management before being able to produce and sell credits.

However, setting a baseline that is more stringent than current practices could come at a cost to point sources trying to reduce their own permit compliance costs. Such a baseline could significantly restrict credit supply by discouraging producers who can supply credits at a relatively low cost but are not meeting the baseline from voluntarily entering a market. A farmer would only be willing to create and sell credits if the expected credit price is high enough to generate the revenue necessary to cover the cost of meeting the baseline *plus* the cost of any measures taken to produce additional abatement. If the price is sufficient for the farmer to enter the market, the amount of abatement equal to baseline is credited toward Load Allocation, while the remainder goes to the market. In aggregate, the sector's credit supply curve shifts to the left, reflecting the higher cost of supplying credits. Higher credit costs will reduce the amount of credits purchased by point sources. The amount of the shift depends on the share of cropland not currently meeting baseline requirements and the amount of abatement farmers must achieve before being eligible to trade. In aggregate, the shift in the sector supply curve increases the market price of credits, reducing the number of nonpoint-source credits purchased by point sources.

There are also consequences from the standpoint of reducing nonpoint-source pollution. If meeting an eligibility baseline discourages those farms that can provide the most abatement at the least cost from participating in a trading program, the overall efficiency of the policy for meeting TMDL goals could be reduced. As demonstrated earlier, traditional design-based policies that might be used to achieve a stewardship baseline consistent with the TMDL are much less efficient than performance-based policies, such as trading.

Stephenson et al. (2010) found that when the eligibility baseline is more stringent than current practices, agricultural credit costs for nitrogen can surpass marginal abatement costs for point sources because the baseline has claimed the lowest-cost pollutant reductions. Ghosh et al. (2011) found that Pennsylvania's baseline requirements significantly increased the cost of entering a trading program, making it unlikely that nonpoint sources that could reduce nutrient losses for the lowest unit costs would enter the market. Wisconsin has expressed concern that U.S. EPA's approach to defining baselines could obstruct agricultural sources' participation in trading programs and possibly impede water quality improvements (Kramer, 2003). The concern is that fewer nonpoint-source credits would be purchased by point sources, and that total abatement costs for regulated sources would be higher than they would be otherwise. None of these studies examined the effect of baseline choice on total nonpoint-source abatement.

We examine this tradeoff between baseline stringency and willingness to trade with the CEAP data from the Chesapeake Bay watershed. We assume that agriculture is not required to meet baseline requirements unless entering into a trading program. We assume that point sources are allowed to seek credits from agricultural sources (consistent with the trading policies of Chesapeake Bay States). Agricultural sources are allowed to sell credits to point sources once a baseline requirement is met. Credits are generated only by the adoption of additional management, even if a field meets the baseline before entering the trading program (no non-additional credits in the market). We select a baseline rule that is based on nitrogen loading rates rather than specific practices, as it is easier to evaluate environmental stringency. This approach is consistent with the baseline rules proposed for Maryland's water quality trading program. We look at loading rate thresholds of 15, 25, 35, 45, and 65 pounds of nitrogen per acre of cropland, along with a baseline based on current practices. We also assume that trading is allowed across sub-basins within the Bay watershed, which is not currently allowed. In our analysis, we examine the relationship between baseline stringency, market equilibrium price, abatement going to the credit market, and abatement going to TMDL Load Allocation.

Demand for Credits in the Chesapeake Bay Watershed

We assume for this analysis that demand for credits from agriculture will come from publicly owned sewage treatment works (POTW). A demand curve for credits is estimated with data from 176 significant wastewater treatment plants in the watershed to represent demand from POTWs (personal communication with John Talberth, Water Resources Institute). Demand is predicated on avoiding costs of treatment upgrades that would be required to meet emission permits under the TMDL. Sources of cost data include Maryland Department of the Environment, Pennsylvania General Assembly Legislative Budget and Finance Joint Committee, and Virginia Department of Environmental Quality.

Treatment costs include capital and operating costs for achieving nitrogen reductions consistent with a Waste Load Allocation for the TMDL. Reductions were calculated as a function of existing flows (3-year average) and both pre- and post-upgrade concentrations. For each plant, costs of reducing nitrogen were modeled on a per-pound basis. After all relevant data for each plant were compiled, a demand curve for credits was plotted (fig. 2). We assume that a credit consists of 1 pound of nitrogen reduction delivered to the tidal waters of the Bay.

Estimating the Supply of Credits

With the data on practice cost, yield, edge-of-field nitrogen loss, and nitrogen delivery to the tidal waters of the bay, we estimate the least-cost option for each observation to supply credits for a trading market, subject to meeting the eligibility baseline. Each observation (considered a representative farm) could add new management systems (nutrient management, cover crops, or water erosion control) to its "current" set of practices, except for those observations/sites already implementing the full suite of management systems. We use the performance-based models described earlier to find the least-cost management choice for each observation. Costs reflect the total costs to the farmer of adopting the practice set (including practice implementation costs), the opportunity costs of changes in nutrient applications and crop yields, and costs associated with increased risk (ignoring any cost-share payments from conservation programs).

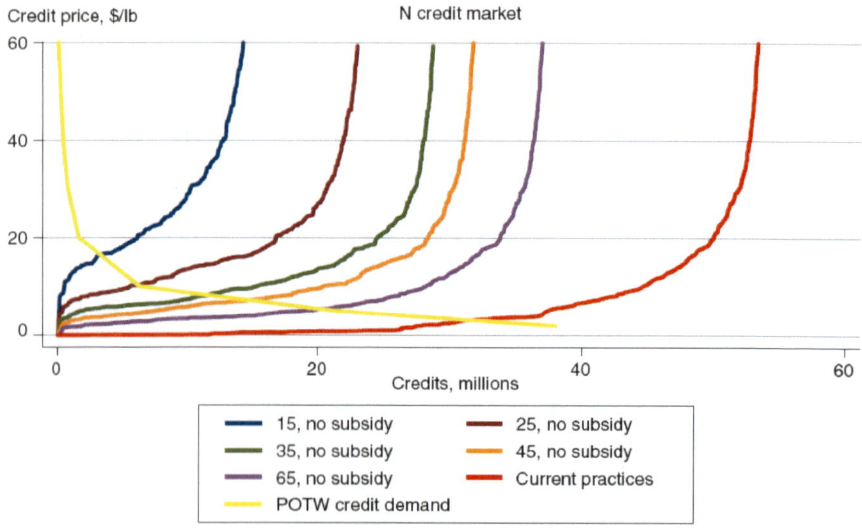

Source: USDA, Economic Research Service analysis of Natural Resources Conservation Service Conservation Effects Assessment Project data.

Figure 2. Nitrogen credit supply and demand curves under alternative loading-rate baselines.

If the observation meets the eligibility baseline, all new abatement goes to the credit market. If the eligibility baseline is not met, abatement is allocated between meeting the baseline (going to the TMDL's Load Allocation) and supplying the credit market. Adopting a single management system often produces enough abatement to meet the baseline and to supply the credit market. The credit price at which it would be profitable for a field to supply credits in the market reflects the cost of achieving the baseline. Aggregating credit supply across all observations for a particular baseline results in a sector credit supply curve for that scenario (fig. 2). We calculate nonpoint-source nitrogen credit supply curves for a "current practices" baseline, and baselines of 65, 45, 35, 25, and 15 pounds of nitrogen loss per acre of cropland. Only about 26 percent of watershed cropland could meet the most stringent baseline before adopting additional practices (table 8). About 85 percent could meet the least stringent baseline.

Table 8. Bay Watershed cropland acres meeting edge-of-field loading limits

Edge of field nitrogen loss (pounds/acre)	Percent of watershed cropland
15	26.5
25	50.8
35	65.4
45	74.2
65	85.3

Source: USDA, ERS analysis of Natural Resources Conservation Service Conservation Effects Assessment Project data.

Results

The "current practices" baseline provides the most credits at lowest cost, as all fields are eligible to supply credits with the implementation of a practice. In this scenario, 31.65 million pounds of nitrogen are sold to POTWs at an equilibrium price of $3.13/lb. (table 9). Point sources realize a benefit of $251 million from being able to purchase nonpoint-source credits rather than installing enhanced treatment technology. All abatement goes to the market and none to Load Allocation. Since all farms can immediately enter the market, most of the credits are coming from fields with high nitrogen losses and without best management systems. Installing practices on such fields produces high amounts of nitrogen reduction for a low unit cost. Fields already under conservation measures cannot provide as much additional abatement and the abatement provided is more costly.

When an N loss eligibility baseline is used to provide more abatement for Load Allocation, supply curves shift to the left, as expected, since some fields cannot sell credits until the baseline is met. Since abatement required to meet the baseline cannot be sold, fewer credits are supplied. The cost of credits supplied reflects the cost of meeting the baseline, plus the cost of any additional measures taken to reduce N losses. The more stringent the baseline, the fewer the available credits (and the further left the supply curve shifts). Under the most stringent baseline (15 lbs. N loss/acre), the equilibrium price in the market would be $16.49 per lb./N, and only 3.29 million credits would be traded (table 9). The benefit to point sources is only $67 million, a decrease of 73 percent from the "current practices" baseline. Some fields not meeting baseline requirements would still find it profitable to meet baseline requirements and sell credits in the market. As such, 3.54 million pounds of N abatement would go toward Load Allocation.

Table 9. Comparison of trading market equilibrium conditions under alternative baselines, with and without practice subsidies

Baseline	Equilibrium price ($/lb. N)	Equilibrium quantity (1 million lbs. N)	Abatement towards ag nonpoint load allocation (1 million lbs. N)	Total ag nonpoint abatement (1 million lbs. N)
Current practices	3.13	31.65	0	31.65
15*, no subsidy	16.49	3.29	3.54	6.83
25*, no subsidy	10.28	6.25	7.49	13.74
35*, no subsidy	8.26	11.62	12.59	24.21
45*, no subsidy	7.23	14.72	18.13	32.85
65*, no subsidy	5.41	20.19	26.42	46.61
15*, subsidized	13.93	4.51	19.84	24.45
25*, subsidized	10.65	6.07	32.18	38.25
35*, subsidized	9.18	8.84	33.08	41.92
45*, subsidized	9.09	9.12	33.94	43.06
65*, subsidized	8.09	12.13	38.54	50.67

*Pounds of nitrogen lost per acre of cropland.

Source: USDA, ERS analysis of Natural Resources Conservation Service Conservation Effects Assessment Project data.

The amount of abatement going to Load Allocation increases as the eligibility baseline is relaxed (table 9). At the least stringent baseline we examined (65 lbs. N/acre), more nonpoint-source abatement (46.6 million lbs. N) is produced at the equilibrium price than under the "current practices" baseline (31.65 million). The reason is that market prices are high enough to induce fields that do not quite meet baseline requirements, but can produce high levels of abatement at low cost, to adopt the necessary practices.

Changing the baseline alters the amount of abatement. If a goal of the program is to maximize total nonpoint-source abatement, then a carefully chosen eligibility baseline could accomplish this. However, the equilibrium credit price in the point/nonpoint trading market is still higher than under the "current practices" baseline and the economic benefits to regulated point sources are reduced.

If a baseline reduces the supply of credits in the market, the government could try to compensate by offering to subsidize practices necessary to meet the baseline. A State may feel that the benefits of trading, in terms of reducing point-source costs and mitigating nonpoint-source pollution, are worth the taxpayer expense of the subsidies. This is currently an accepted use of conservation program funds (American Farmland Trust, 2013).

We examined this policy by calculating nonpoint-source nitrogen credit supply curves when the costs of meeting the baseline are not paid by the farmer. For simplicity, we assume a 100-percent cost-share. We also assume that abatement generated by the subsidized practices is not eligible for sale as credits, even if the baseline is exceeded (consistent with rules against so-called double dipping, or receiving multiple payments for the same action). All abatement from subsidized practices goes to the nonpoint-source Load Allocation of the TMDL. Credits would be produced for practices adopted once the baseline is met.

For all but the most stringent baseline (15 lbs. N/acre), the provision of a subsidy actually reduces the number of credits going to the market (table 9, fig. 3). The reason for this counterintuitive result is that abatement generated by a practice cost-shared by the government cannot be sold as credits in the market, even if the eligibility baseline is exceeded. As a result, some of the credits that were available to the market when the subsidy was not offered instead go to Load Allocation; the supply curve for that stringency level shifts to the left with the subsidy. Point sources do not benefit from the subsidy, but total nonpoint-source abatement is higher with the subsidy for all baselines. If the subsidy is less than 100 percent (as are cost-shares in most conservation programs), the incentive would be reduced and less nonpoint-source abatement would be generated.

These results are driven in part by the "lumpiness" of our management options and the limited number of management options considered. Adopting a practice may produce a large amount of abatement that cannot be "fine-tuned" to exactly match demand. With more management choices, a farmer would have greater control over how much abatement can be provided and the impact of the subsidy on credit supply would probably be reduced; more abatement would go to the market and less to Load Allocation. A baseline based on management practices rather than nitrogen loadings would not face this issue.

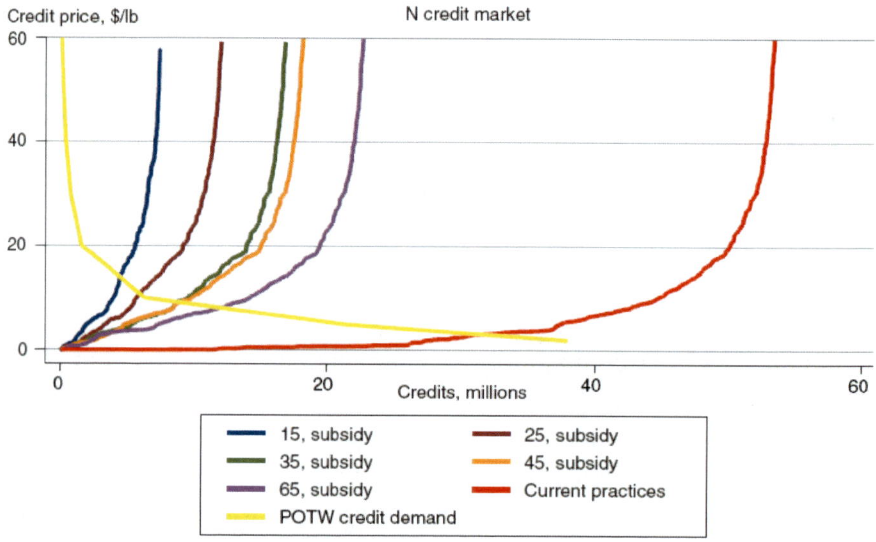

Source: USDA, Economic Research Service analysis of Natural Resources Conservation Service Conservation Effects Assessment Project data.

Figure 3. Nitrogen credit supply and demand curves in the presence of program subsidies under different loading-rate baselines.

Discussion

The stewardship-based eligibility baselines for point/nonpoint trading could restrict the number of credits that would be supplied in a trading program. Existing baseline rules are intended to encourage achievement of the

TMDL Load Allocation for agriculture. However, the rules may unnecessarily constrain markets, increasing credit prices and reducing potential efficiency gains.

Stringent nitrogen loading baselines did not provide much "extra" abatement for Load Allocation, while they significantly reduce the volume of credits traded. Relaxing the baseline increased abatement for both the market (benefiting point sources) and for Load Allocation. Offering a subsidy to cover the cost of meeting the baseline significantly increased total abatement. In our analysis, much of this increase went to Load Allocation. With more choices of conservation practices than we considered and different rules for how credits are awarded, a subsidy could provide more credits for the market. Our findings emphasize the importance of carefully evaluating these factors when designing a trading program.

This analysis does not consider market design features such as trading ratios and geographic limits on markets that would further restrict trading and increase credit prices. Given that impediments to trading are many (Ribaudo and Gottlieb, 2011; King and Kuch, 2003; Woodward and Kaiser, 2002; Shortle, 2013) and the stated need to reduce the costs facing regulated point sources for meeting the TMDL, baseline choice has a major influence on whether point sources can reduce these costs.

The worst nonpoint polluters tend to have the largest range of abatement possibilities, so the most effective abatement program of any kind (trading or otherwise) incentivizes these farms to participate. Setting stringent eligibility baselines tends to reduce participation in trading by this group, because of the high cost of meeting baseline requirements. Without a regulatory means of forcing nonpoint-source producers to adopt management practices consistent with the Load Allocation of the TMDL, the consequences of choosing an eligibility baseline are magnified.

MANURE MANAGEMENT AND THE TMDL

Animal agriculture is a significant part of the agricultural economy in the Chesapeake Bay watershed. Of the 83,800 farms in the watershed in 2007, animal farms with predominantly confined animals accounted for about 13,800 operations (16 percent), with a combined average daily inventory of roughly 2.0 billion pounds of poultry, dairy, swine, and feedlot beef. Animal agriculture is also a significant source of nutrient loadings to the Bay. Animal operations produce roughly 99,400 tons of recoverable manure nitrogen and

44,200 tons of recoverable manure phosphorus annually (USDA, NRCS, 2012), most of which is applied to available cropland on the source operation. However, confined animal feeding operations often produce more manure nutrients than can be used by crops grown on the land managed by the operation (Ribaudo et al., 2003; Gollehon et al., 2001). This has led to problematic storage of surplus manure and overapplication of manure nutrients, with resulting nutrient losses to the environment (Ribaudo et al., 2003; Aillery et al., 2005). An estimated 17 percent of the nitrogen entering the Bay and 26 percent of the phosphorus has been attributed to animal operations (NRC, 2011).

Confined animal feeding operations are regulated to control for water-quality impacts under the Clean Water Act. The largest animal feeding operations (Concentrated Animal Feeding Operations or CAFOs) require NPDES permits if discharge from the housing facilities enters surface waters through a pipe or ditch. In addition, all CAFOs are expected to have a nutrient management plan that balances the application of manure nutrients to cropland controlled by the operation with the agronomic needs of receiving crops. Most of the Bay States have requirements on the storage and application of manure nutrients that cover all sizes of operations, not just the largest. Maryland and Delaware also pay animal operations to haul manure out of their States.

Despite existing Federal and State programs intended to encourage sound manure management, data from the 2007 Census of Agriculture indicate a significant imbalance between manure nutrients produced in the watershed and the nutrient uptake potential of the crops receiving manure. Even if confined animal operations in the CBW fully utilized the crop and pasture land under their control for manure application, only about 47 percent of the manure nitrogen produced could be assimilated onfarm.[5] Clearly, responsible nutrient management requires moving significant quantities of manure off animal operations for abatement practices to be effective. State Watershed Implementation Plans acknowledge this continued imbalance by setting reduction goals for nutrient emissions from manure applications. The TMDL creates the impetus to make the fullest use possible of these requirements.

Our earlier analysis of improved management on crop acres covered all cropland, including that owned or leased by animal operations. However, the CEAP data did not allow us to single out this cropland. Because animal operations have a continuing, self-supplied source of crop nutrients (animal waste) that cannot be easily "turned off," the management problems they face are different than those of crop producers who purchase all their nutrients. The cost of moving excess manure nutrients to additional cropland where it can be applied efficiently, or to other uses, adds to the overall cost of improving nutrient management on animal operations (Ribaudo et al., 2003). To examine these costs, we apply a different data set and model.

Much of the watershed's excess manure is concentrated in major poultry-producing areas of southern Maryland and Delaware, as well as northwest Virginia and eastern West Virginia. Concentrated dairy production across south-central Pennsylvania adds to the manure excess (fig. 4). Balancing nutrient supply and demand within the watershed will require moving manure from areas of surplus to areas where it can be used without posing a risk to water quality. Cropland receiving commercial fertilizer can use manure nutrients to meet all or some of the crops' nutrient needs, but moving manure from animal operations requires cropland managers willing to use it following good nutrient management practices. Alternative uses for manure, such as energy and commercial fertilizer production, could also absorb some of the regional excess. Manure (nutrients) may also be transported out of the watershed entirely. All of these options require hauling manure from farms with excess, which can be costly (Ribaudo et al., 2003).

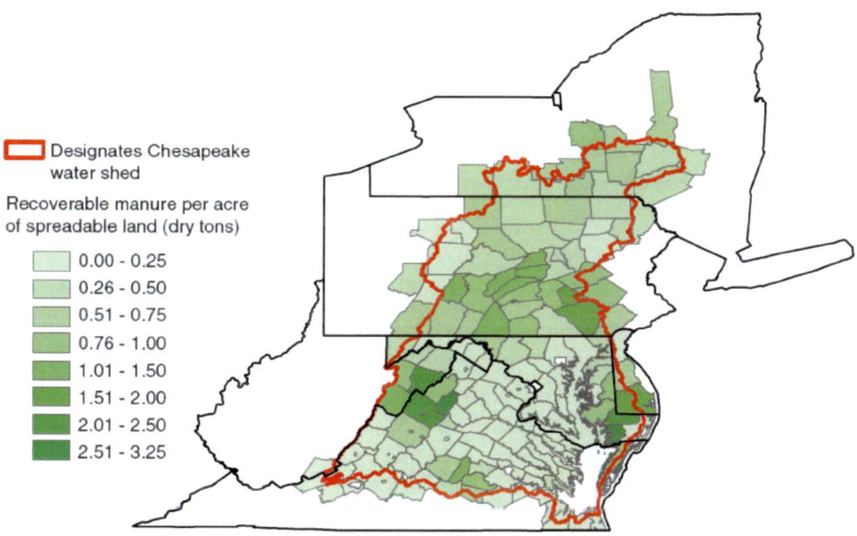

Source: USDA, Economic Research Service analysis of 2007 Census of Agriculture data.

Figure 4. Total recoverable manure (dry tons) per acre of spreadable land (all cropland plus half of permanent pasture) in the Chesapeake Bay watershed, by county, 2007.

The costs associated with managing surplus manure nutrients depend not only on individual farm conditions but also on the *interaction* among animal operations, and their proximity to available land for manure spreading. Our analysis uses the Chesapeake Bay Regional Manure Transport Model, which accounts for the competition for available cropland land among animal producers in the Chesapeake Bay Watershed (see Aillery et al., 2005, for a complete description of the model). We assume that all animal feeding operations (AFOs) are moving toward meeting the nutrient management goals laid out in the State Watershed Implementation Plans. The model and its results reflect a regional planning perspective emphasizing the cost determinants and feasibility of different manure management strategies at the watershed scale.

Modeling Manure Management in the Chesapeake Bay Watershed

In areas of the Chesapeake Bay watershed where confined animal production is concentrated, implementation of manure management policies poses tremendous challenges. If the manure produced exceeds potential local use, producers may choose to (1) transport the manure ever greater distances until enough land can be found for application or (2) pursue technologies that either utilize manure as a production input or transform the manure to a value-added product that is more readily transportable and usable. Changes in feed management, such as adding phytase to enhance phosphorus utilization for non-ruminants (poultry, hogs), can help reduce nutrients excreted in manure to some extent.[6] Manure amendments, such as alum in poultry litter, have also been used to lessen phosphorus runoff from field-applied manure.[7] Beyond this, the only recourse is to reduce the number of animals in the watershed. Florida, for example, reduced its dairy numbers via a Dairy Buyout Program to reduce nutrient runoff from dairy farms in the Lake Okeechobee watershed to protect the Everglades (Schmitz et al., 1995).

The Chesapeake Bay Regional Manure Transport Model helps us account for landowner willingness to apply manure and non-cropland manure uses. Our model is designed to minimize the total regional costs of manure management, transport, and application for use on agricultural lands in the CBW, given the existing scale and structure of the animal industry and manure storage/hauling technologies currently in use. The regional specification captures the element of competition by modeling access to available farmland, simulating within-county and out-of-county manure flows based on manure nutrient content and assimilative capacity of receiving farmland, and computing the associated hauling and application costs. Explicit modeling of competition for land on which to spread manure differentiates the model from existing farm-level models.

The model was developed to (1) provide a mechanism to track manure and related nutrient flows within the CBW, from confined animal farms to offsite application and use; (2) compute the regional costs of applying manure to land, given manure transfers dictated by nutrient flows and manure hauling weight; and (3) provide a framework for evaluating proposed land-application regulations and alternative nutrient management policies. The model was first used to evaluate the impacts of new Clean Water Act regulations on CAFOs (Ribaudo et al., 2003). For this analysis, several changes were made to accommodate more recent data from the 2007 Census of Agriculture and a revised methodology for estimating hauling distance (see Appendix, "Updates to the Chesapeake Bay Regional Manure Transport Model").

The county is the primary modeling unit. The county-level specification provides consistency with census of agriculture data and other data sources and permits differentiation of institutions and regulatory conditions across county and State political boundaries within the watershed. County and other data are used to capture heterogeneity in production, technology use, and land resource availability across the region, though our model may not represent the conditions on any particular farm.

Model output reported at the 6-digit Hydrologic Unit Codes (HUC) is used to highlight variation in production conditions and policy response at a sub-basin scale.

Study Scenarios

The cost of achieving TMDL pollutant abatement goals for animal agriculture will depend on various factors that influence the production and disposition of excess manure-nutrients in the CBW. Using the Chesapeake Bay Regional Manure Transport Model, we first simulate onfarm AFO manure excess and off-farm manure use on non-AFO crop farms under conditions reflected in the 2007 Census of Agriculture. The Base Case approximates the "current" costs of manure handling based on reported off-farm manure use in 2007, given that accurate estimates of the share of applied manure meeting nutrient standards are unavailable. We then evaluate four policy scenarios: (1) compliance with a nutrient management plan across differing levels of willingness of non-AFO crop farms to accept manure; (2) compliance with a more restrictive requirement for nutrient management; (3) the presence of manure demand for non-cropland uses (energy generation), and (4) reductions in the scale of the confined animal sector (see table 10 for scenario labels).

Base Case

The *Base2007* simulation was constructed to approximate the disposition of manure and manure handling costs under current (2007) conditions, providing a context for examining changes in manure handling under alternative NMP compliance assumptions. The base case fixes onfarm manure use (AFOs) and off-farm manure use (non-AFO crop farms) to reported levels of acreage receiving manure in the 2007 Census.[8] Onfarm manure excess was calculated from estimated recoverable manure nutrients produced on AFOs and crop-nutrient uptake on AFO cropland and pastureland in 2007, assuming an N standard for land application of manure and adjustments in manure-ready acreage due to P-limited soils (Kellogg et al, forthcoming). In the base case, we assume that all manure hauled off the farm is applied at agronomic rates under an N application-touptake ratio of 1.4 (N efficiency of 0.71), reflecting historically acceptable manure N losses (USDA, NRCS, 2011). Onfarm manure accounts for all remaining manure—including excess that is not applied to land off the farm and not otherwise removed for non-cropland uses.

Under the *Base2007* simulation, onfarm manure applied on AFOs accounts for a significant share (69 percent) of total manure produced across Chesapeake Bay counties (fig. 5). Onfarm manure includes 3.5 million dry tons (56 percent of total) sufficient to meet the agronomic requirements of crops produced on AFO farms, plus an additional 0.8 million tons (13 percent) that is assumed to be overapplied on AFO farms. Such overapplications are consistent with previous research (Ribaudo et al., 2003; Gollehon et al., 2001), and align closely with NRCS simulation assumptions for onfarm AFO manure use in 2007 (Kellogg, et al., forthcoming).

Manure moved off the farm and applied to cropland accounts for 1.9 million tons, or 31 percent of total manure produced. Approximately 1.1 million tons of manure moved off-farm is applied within the source county, while 0.8 million tons are hauled to other counties within the watershed. A small share of manure is diverted to non-cropland uses—primarily for commercial fertilizer products and small-scale composting of manure sold for non-cropland uses within and outside the basin. We estimate this amount at roughly 1 percent of total manure production in the Chesapeake Basin counties.[9]

Table 10. Study simulations

Base case—2007	BASE
Nutrient Management Plan compliance scenarios	
Willingness to accept manure	
30%	WTAM30
50%	WTAM50
70%	WTAM70
90%	WTAM90
Full conservation treatment	HICON50
Manure for biomass energy	
5% M excess	BIOMS05
10% M excess	BIOMS10
15% M excess	BIOMS15
20% M excess	BIOMS20
Animal-unit reductions	
2% AUs	AURED02
4% AUs	AURED04
6% AUs	AURED06

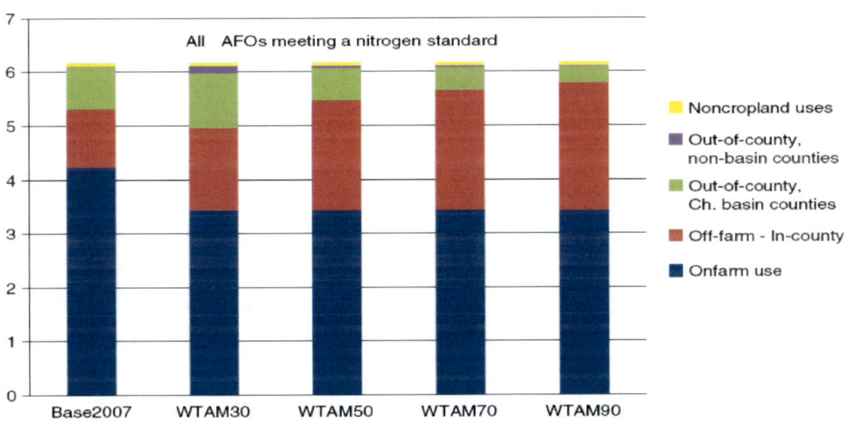

AFO = Animal feeding operation; NMP = Nutrient management plan; WTAM = Willingness to accept manure.

Source: USDA, Economic Research Service analysis of 2007 Census of Agriculture data.

Figure 5. Disposition of manure in the Chesapeake Bay watershed under current conditions and full NMP compliance, by share of non-AFO crop producers willing to accept manure.

Manure hauling costs, which are expected to increase with full NMP implementation, are an important concern for the animal industry (Ribaudo et al., 2003). Under the *Base2007* simulation, annual manure hauling costs are estimated at $171 million across basin counties, with on-farm hauling costs of $116 million accounting for a significant share.[10] Off-farm hauling costs of $55 million were divided between in-county ($34 million) and out-of-county ($22 million) hauls. Poultry litter accounts for a large share of long-haul manure transfers, with wetter slurry and lagoon waste more typical of dairy and hog operations tending to shorter within-county hauls.

Manure was applied on approximately 1.6 million acres across the Chesapeake Bay watershed in 2007, including 1.1 million acres on AFOs and 0.5 million acres on non-AFOs (altogether 36 percent of the 4.4 million cropland acres in the CBW) (USDA, NRCS, 2011). An estimated 12 percent of total farmland received manure in the watershed, with acreage shares varying considerably across sub-regions. County-level estimates of manure-nutrient assimilative capacity suggest considerable opportunity to reallocate surplus manure within the watershed.[11]

Nutrient Management Plan Compliance Scenarios

In the following scenario analyses, we assume that all AFOs and non-AFOs in the watershed apply manure under an approved nutrient management plan (NMP) that defines acceptable application rates, timing, and application methods (consistent with nutrient management described earlier and with rules developed by the Bay watershed States). (While all manure applied to cultivated cropland is assumed to be incorporated or injected, the full suite of conservation treatments may not necessarily be in place to minimize nutrient losses from applied manure). As more AFOs comply with manure-nutrient standards under approved plans, manure applied onfarm is reduced and a greater share will have to be moved off the farm. Increased competition for land to spread manure will, in turn, increase the distance required to access suitable land. Higher manure hauling costs faced by animal producers is an important policy challenge in moving toward full NMP compliance.

Willingness to Accept Manure (WTAM)

The willingness of crop farms to accept manure will be a key determinant of future hauling costs. A reluctance to apply manure may reflect various concerns, including potential odor in nearby residential areas, manure storage requirements, variability in manure-nutrient content, potential dispersion of weed seeds and pathogens, and possible litigation from nearby nonfarm housing. Manure rotational requirements on soils with high phosphorus content may also limit applied manure in some areas. The willingness of landowners to use manure affects manure handling costs through changes in both the extent of suitable land and the hauling distance required to access available acreage.

Under our NMP compliance scenarios, we assume that all AFO farms are willing to use their own manure onfarm to meet crop nutrient requirements. For non-AFOs, we apply four WTAM levels to manure-ready acreage across the CBW—30, 50, 70, and 90 percent. The range of WTAM levels reflects the uncertainty surrounding landowner attitudes to manure, and the sensitivity of hauling costs to alternative WTAM specifications. While WTAM assumptions are applied uniformly across the watershed, we recognize that rates may vary across sub-basins. For example, landowner willingness to use manure may be higher in animal-producing areas with an established tradition of manure use and more active local manure markets.

Under the NMP-compliance scenarios, onfarm manure use accounts for about 56 percent of manure produced in the CBW counties—or that amount required to cover nitrogen requirements of crops produced on AFOs (assuming an N application-to-loss ratio of 1.4). The decline in onfarm manure use from *Base2007* levels is mirrored in an expansion in off-farm manure transfers (fig. 5). With increasing willingness of non-AFO crop producers to apply manure, a greater share of manure transfers are used within the source county, with lesser shares hauled to other counties both within and outside the basin.

As the region moves toward full NMP compliance, off-farm manure hauling and application costs are projected to expand relative to *BASE2007* conditions (fig. 6). Increased costs reflect both expanded acreage receiving manure and increased hauling costs required to access additional acres. Off-farm hauling costs—estimated at $55 million basinwide under *BASE2007*—increases by a range of $15 to $27 million, depending on assumptions about crop producers' acceptance of manure. Increasing the willingness to accept manure from 30 percent to 90 percent reduces hauling costs (basinwide) by about 15 percent. Off-farm land application (and incorporation) costs increase basinwide from $11 million to roughly $14 million, reflecting the expansion in manure spreading requirements.

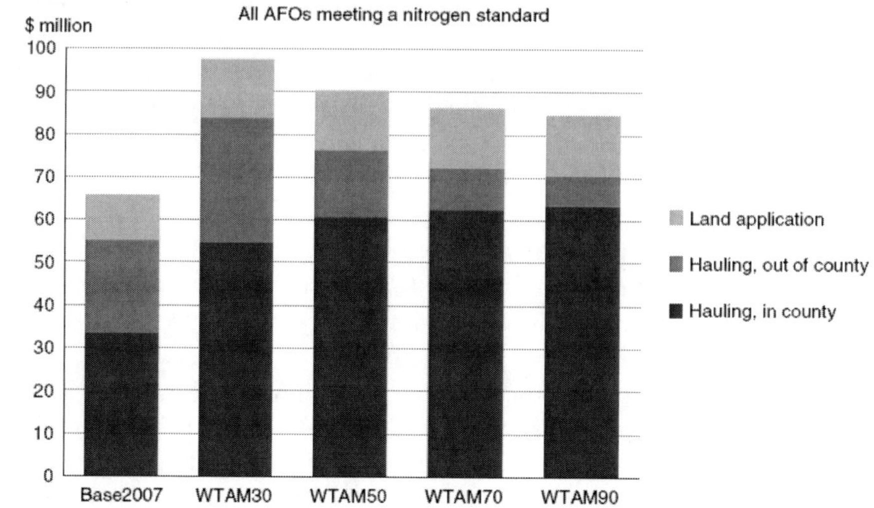

AFO = Animal feeding operation; NMP = Nutrient management plan; WTAM = Willingness to accept manure.

Source: USDA, Economic Research Service analysis of 2007 Census of Agriculture data.

Figure 6. Off-farm manure hauling and application costs under current conditions and full NMP compliance, by share of non-AFO crop producers willing to accept manure.

WTAM and Intra-Regional Considerations

The effect of WTAM on off-farm hauling costs varies widely across the watershed, reflecting considerable heterogeneity in production and land resources. In general, average off-farm hauling distance declines as non-AFO willingness to accept manure increases (table 11). This is most marked in the Potomac and Upper Chesapeake sub-basins where poultry production is concentrated. In areas where animal production is less concentrated, as in the Upper and West Branch Susquehanna basins, increases in manure acceptance have a negligible impact on hauling distance in absolute terms. The sensitivity of hauling distance to WTAM is further influenced by cropland density (fig. 7). In the James and Lower Chesapeake sub-basins, for example, where animal concentrations are comparatively low, the relative sparseness of contiguous cropland results in longer average hauls for a given WTAM. In all cases, the impact of a shift in manure acceptance depends on current WTAM levels. Where manure acceptance is lower (e.g., 30 percent), increased rates would help relieve competition for manure-ready land; the effect is lessened when manure acceptance is initially higher.

Average hauling cost per dry ton of manure has a somewhat inverse relationship with hauling distance (table 12). Using *WTAM50* as a benchmark scenario,[12] average hauling costs range from $40/ton to $52/ton across the Susquehanna sub-basins where dairy production (with relatively high-moisture manure content) is predominant. In contrast, in the Upper Chesapeake and Potomac sub-basins, where much of the (comparatively dry) poultry litter is produced, hauling costs range from $17/ton to $27/ton. Average hauling costs per dry ton are generally less sensitive to WTAM levels than average hauling distance, given that manure hauling costs are based on both a cost-per-ton and cost-per-unit-distance.

Table 11. Average off-farm hauling distance, by willingness-to-accept manure (WTAM) and sub-basin (HUC-6)

		WTAM30	WTAM50	WTAM70	WTAM90
Sub-basin			*Miles*		
HUC_20501	Upper Susquehanna	2.0	1.7	1.5	1.3
HUC_20502	West Branch Susquehanna	2.4	2.0	1.7	1.6
HUC_20503	Lower Susquehanna	13.2	4.4	4.0	3.1
HUC_20600	Upper Chesapeake	28.5	16.7	8.9	7.6
HUC_20700	Potomac	43.0	34.2	22.8	17.0
HUC_20801	Lower Chesapeake	25.4	18.2	14.2	11.7
HUC_20802	James	23.5	14.1	11.0	10.3

Source: USDA, Economic Research Service analysis of 2007 Census of Agriculture data.

Under scenario *WTAM50*, several manure-nutrient surplus counties of the Upper Chesapeake, Potomac, and Lower Susquehanna basins show significant net reductions in manure (fig. 8), as manure exports to nutrient-deficit areas exceed county inflows. Net imports of manure are positive for portions of the south-central CBW (James and Lower Chesapeake sub-basins) and northern CBW (Upper and West Branch Susquehanna sub-basins). However, much of the onfarm manure excess produced in the CBW remains within the sub-basin of origin, with imports often concentrated in receiving counties adjacent to high-surplus counties.

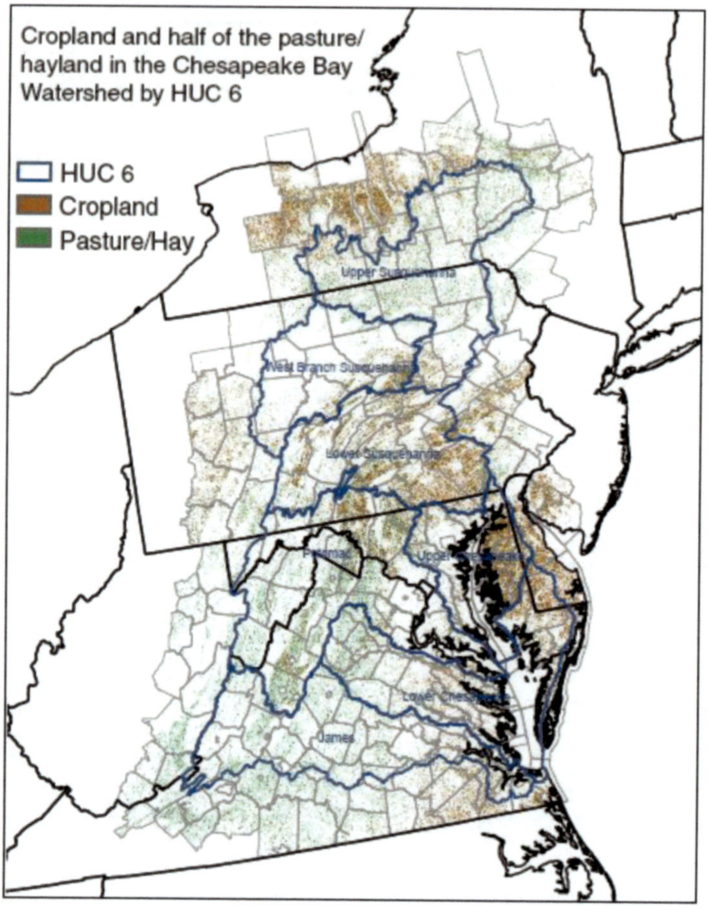

Source: USDA, Economic Research Service analysis of 2007 Census of Agriculture data.

Figure 7. Cropland and pastureland area available for manure spreading, with Chesapeake Bay sub-basin (HUC6) delineations, 2007.

Full Conservation Treatment

While NMP compliance scenarios reflect generally high levels of manure management, further reductions in manure-nutrient losses may be achieved through the adoption of additional conservation treatments appropriate for a given field. Increasing nitrogen use efficiency from 0.71 to 0.83 (N application-to-uptake ratio of 1.2 or less) may require reduced application rates, as well as such practices as cover crops, filter strips and riparian buffers that capture excess nutrients and reduce pollutant runoff. Somewhat less apparent are the implications of moving toward full conservation treatments on manure handling costs in the watershed.

Table 12. Average off-farm hauling cost, by willingness-to-accept manure (WTAM) and sub-basin (HUC-6)

		WTAM30	WTAM50	WTAM70	WTAM90
Sub-basin		*$/dry ton manure hauled*			
HUC_20501	Upper Susquehanna	52.0	52.0	51.9	52.7
HUC_20502	West Branch Susquehanna	44.7	44.4	44.4	45.0
HUC_20503	Lower Susquehanna	43.3	39.9	39.9	39.8
HUC_20600	Upper Chesapeake	19.8	16.7	14.3	12.9
HUC_20700	Potomac	30.3	27.3	24.7	22.1
HUC_20801	Lower Chesapeake	21.6	20.0	19.3	17.0
HUC_20802	James	28.4	25.5	25.0	23.2

Source: USDA, Economic Research Service analysis of 2007 Census of Agriculture data.

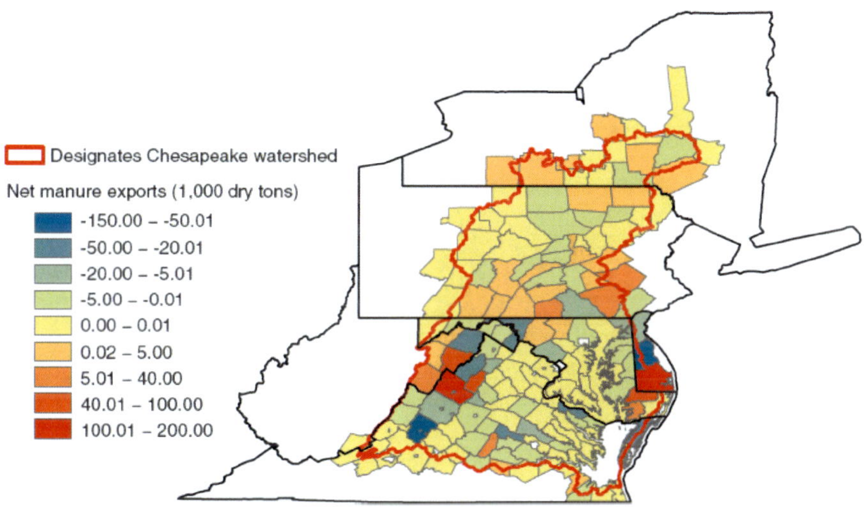

NMP = Nutrient management plan.

Source: USDA, Economic Research Service analysis of 2007 Census of Agriculture data.

Figure 8. Net manure exports (1,000 dry tons) by Chesapeake Bay Watershed county, assuming full NMP compliance and 50 percent willingness to accept manure.

Figure 9 shows manure hauling and application costs under two cases: NMP compliance, with and without full conservation treatments (*HICON50* and *WTAM50*). Full conservation treatments increase application costs by $2 million (15 percent) due to expanded acreage receiving manure.

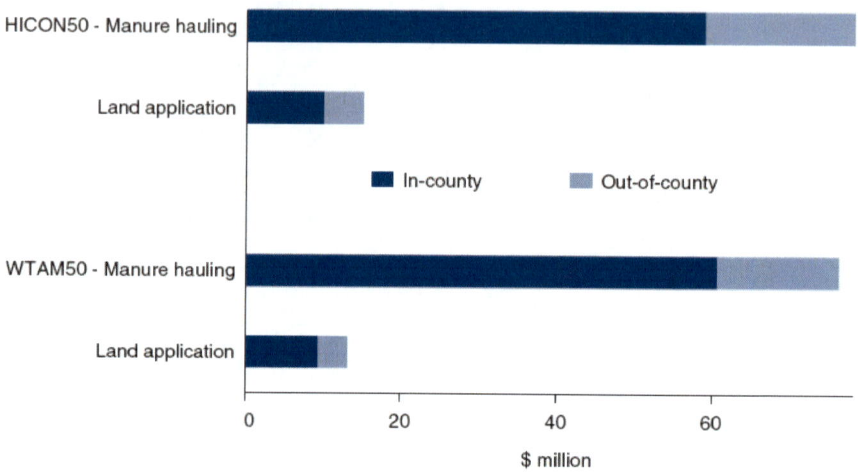

NMP = Nutrient management plan; WTAM = Willingness to accept manure; HICON = Full conservation treatment assuming a willingness to accept manure of 50 percent.

Source: USDA, Economic Research Service analysis of 2007 Census of Agriculture data.

Figure 9. Off-farm manure hauling and application costs in the Chesapeake Bay watershed under full NMP compliance and 50 percent willingness to accept manure, with and without full conservation treatments.

Hauling costs required to access additional acres are also increased by $2 million (2 percent).[13] Higher manure handling costs are likely shared by AFO and non-AFO crop farms, given the dual effect on manure hauling and land application costs.

Manure Use for Biomass Energy

The use of manure as a biomass feedstock for electricity power has been promoted as an innovative means to address regional goals for management of excess manure and renewable energy generation (Ribaudo et al., 2003). While a number of concept proposals have been advanced in the Chesapeake Bay region, uncertainty in energy markets, air-quality concerns, ash disposal issues, and other local siting issues have generally limited their support. The impact of a large-scale biomass facility on local manure supplies has been a particular concern of the agricultural community (Pease et al., 2012).

Here we evaluate the effect of manure use for energy generation on AFO off-farm hauling costs for land-applied manure, and potential implications for the crop sector. Against our benchmark *WTAM50* scenario, we consider five levels of regional manure feedstock demand for power generation—50,000, 100,000, 150,000, 200,000, and 250,000 dry tons (out of 6.2 million dry tons of excess produced) – based on proposed energy projects. Manure feedstock is assumed to be drawn from poultry litter and other semi-dry manure sources (30 and 50 percent moisture content, respectively). As we are not evaluating a specific sited facility, reductions for biomass feedstock are applied uniformly across all excess dry and semi-dry manure sources from across the watershed to give total dry-weight tonnage. Base assumptions for manure used in pelletization and small-scale composting are held constant.

Figure 10 shows the reduction in (1) off-farm hauling costs and (2) value of chemical fertilizer offsets under alternative levels of manure diversions for power generation. With reductions in manure excess to supply demand for biomass feedstock, the decline in off-farm hauling costs for land-applied manure range from $1.1 to $5.0 million basinwide. Reductions in hauling cost moderate with greater manure diversions, as the most costly hauls are retired first. To the extent that manure transport costs are borne by the animal operations, cost savings accrue largely to producers in the Potomac and Upper Chesapeake sub-basins where, not surprisingly, much of the interest in a facility siting has focused.

With reduced manure excess, and resulting declines in acreage receiving manure, the value of forgone manure—based on the cost of equivalent levels of chemical N and P fertilizer—ranges from $1.7 to $8.4 million. The response function is nearly linear across manure diversions, as nutrient uptake rates averaged across affected acreage basinwide remain fairly constant (fig. 10). A tightening of available manure supplies may result in additional costs for area crop producers who have come to rely on manure, to the extent that the full value of applied manure exceeds purchase costs. Under the *BIOMS250* scenario, the 250,000 dry tons of manure diverted for energy generation results in 76,800 fewer non-AFO acres using manure across the basin (relative to the zero-biomass case). The forgone value of poultry litter as a manure source—based on equivalent levels of chemical N and P—is estimated at $33/dry-ton.[14]

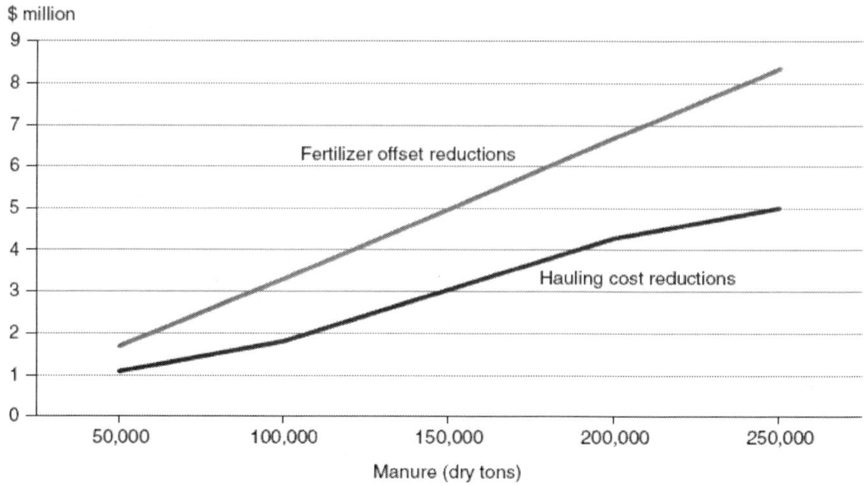

Source: USDA, Economic Research Service analysis of 2007 Census of Agriculture data.

Figure 10. Effects on the value of fertilizer offsets and off-farm hauling costs of the use of manure as in input to biomass energy production in the Chesapeake Bay watershed, assuming full Nutrient Management Plan compliance and 50 percent willingness to accept manure.

According to Pease et al. (2012), brokers in the Shenandoah Valley pay as much as $12 to $15 per ton of poultry litter, or roughly $50 per dry-ton of manure when adjusted for moisture (30 percent) and bedding (10 percent, wet weight) content. Higher purchase prices may reflect, in part, the value of additional nutrients beyond N and P, as well as soil-enhancing benefits of organic material. While manure charges paid by crop producers are not well documented and returns to applied manure would vary across the basin, high litter prices and prospects of rising manure demand for biomass energy suggest at least the potential for higher costs to the regional crop sector.

Scale Reductions in the Animal Sector

The challenge of managing manure nutrients in the Chesapeake Bay watershed is essentially one of nutrient flows and balance—large imports of animal-feed nutrients supporting an increasingly concentrated animal industry, and only a limited amount of available cropland to assimilate excess manure nutrients. Where excess nutrients cannot feasibly or economically be addressed through dietary regimes, improved management practices, increased landowner acceptance of manure, or expansion of non-cropland uses— all of which must be part of an integrated solution—reductions in the scale of animal production may be viewed as a reasonable approach. The Florida buyout of dairy operations in the northern Everglades basin suggests that voluntary, compensated reductions in animal herd size to improve water quality and environmental resources may represent a viable policy option.

Here we consider the effect of reducing the scale of the animal sector in the Chesapeake Bay watershed on (1) off-farm manure hauling costs and (2) returns (above operating costs) to animal production. Using *WTAM50* scenario as a benchmark, we consider three levels of reductions in manure-N excess—5, 10, and 15 percent—roughly equivalent to reductions of 2, 4, and 6 percent of animalunits[15] basinwide. Excess manure is removed on a uniform percentage basis across all manure types in the watershed, and reductions in animal-units are assumed to be applicable only where AFO manure exceeds onfarm crop requirements.

Figure 11 shows the effect of a reduction in animals on manure hauling costs and returns (above operating expenses[16]) to the animal industry. Basinwide, a 2-percent reduction in animal-units results in a loss of approximately $52 million to the animal sector, measured as weighted net returns across animal species types. Reducing nutrient excess by 15 percent would require roughly 6 percent fewer animal-units in the basin, valued at more than $150 million. The number and total value of animal-units forgone to meet manure-nutrient excess targets would vary over sub-basins, reflecting both the concentration of excess manure and the species composition of animals.

The potential savings in off-farm hauling costs—ranging from $16 to $44 million depending on the level of nutrient excess targeted—are well below the forgone returns to animal production. Moreover, as forgone returns increase linearly with higher targeted levels of excess, the increase in sector savings decline as less costly hauls are eliminated. While operators facing exceptionally high manure hauling distances would have increased incentive to participate in a voluntary herd/flock reduction program, savings in manure hauling costs alone may not be a significant driver for many operators.

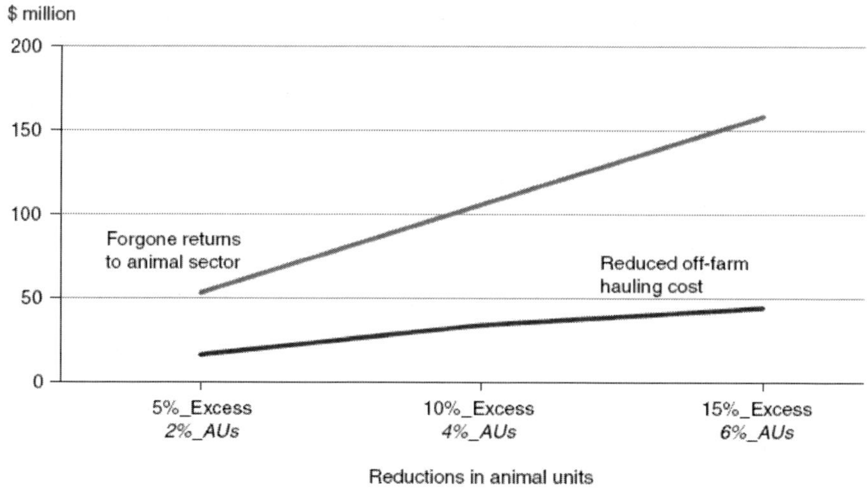

Source: USDA, Economic Research Service analysis of 2007 Census of Agriculture data.

Figure 11. Effects on off-farm hauling costs and forgone returns to the animal sector from reducing excess manure through a reduction in animal units in the Chesapeake Bay Watershed.

CONCLUSION

The animal sector faces increasing costs for off-farm manure hauling as it moves toward full NMP compliance for manure nutrients—both in terms of share of manure managed under approved plans and level of conservation treatments. The magnitude of these additional costs (about $18 million assuming nutrient management and 50 percent of farmers willing to accept manure; $22 million assuming full conservation treatment and the same WTAM) is much smaller than the costs estimated for the design-based policy scenarios that require treatment on large shares of cropland ($337 million to $1.2 billion). Increased hauling costs, while important to animal operations, appear to be a relatively small portion of the total costs to the agriculture sector of meeting TMDL goals.

Our analysis suggests these costs may be potentially mitigated through incentives to increase the willingness of landowners to accept manure. This might involve technical assistance for NMP development and cost-sharing for improved manure storage and handling technologies. The development of local markets for manure transfers, already occurring in the CBW (Pease et al.,

2012), may further promote application of manure on land. Targeting areas where competition for manure-suitable land is greatest and landowner willingness to accept manure is relatively low may achieve the greatest benefits.

Costs may also be mitigated through point/nonpoint water quality trading, as for cropland. For those animal operations not covered by Federal or State nutrient management requirements, or those operations that have already met their nutrient management requirements, moving manure off the farm as part of a nutrient management plan and applying it to cropland elsewhere under an approved plan could be a way to create credits for sale in a market. However, confined animal operations are likely to be at a competitive disadvantage relative to crop producers, who do not incur the manure hauling costs (Sneeringer, 2013). Any decrease in the application of manure nutrients on cropland controlled by an animal operation necessitates an increase in the amount of manure moved elsewhere. In contrast, reducing nutrient application rates on a crop farm to generate credits actually reduces fertilizer costs.

Biomass energy generation has been suggested as a potential strategy for management of excess manure. If biomass facilities were to successfully address financial and environmental concerns raised with earlier proposals, increased demand for manure as a biomass feedstock could reduce manure hauling costs while increasing prices paid to animal producers. Siting of plant facilities would be important to maximize access to available manure supplies. However, manure diverted for energy expansion could impose costs on the regional crop sector if producers who had been using manure were required to rely more exclusively on commercial fertilizer.

Reducing the number of animals that contribute to manure-nutrient excess represents perhaps the most direct response to addressing concerns of excess manure. However, a voluntary animal buyout program is likely to be quite expensive and would be inconsistent with the watershed States' stated goal of preserving agriculture in the watershed. Moreover, our simple analysis does not consider the broader costs of restructuring and potentially relocating animal production to other regions. But it does provide some insight on the magnitude of costs that may be involved in downscaling the animal sector to meet regional water-quality goals.

Policy Considerations for Meeting Nonpoint-Source Abatement Goals

The Chesapeake Bay TMDL was implemented largely in response to the failure of conventional policies to adequately address nonpoint-source pollution. The TMDL is forcing States to reconsider how to deal with agriculture and other nonpoint pollution sources. While there is currently a heavy reliance on traditional, voluntary approaches, a number of alternative policy instruments could be applied to promote improved nutrient management. In addition, a number of choices in how a particular policy, including traditional voluntary approaches, is designed could also improve cost-effectiveness.

Goals Can Be Achieved by Addressing Relatively Little Crop Acreage

The Optimal scenario shows the least-cost allocation across sub-basins of the three conservation systems we allowed farms to adopt to achieve pollution abatement goals. Results indicate that the goals for agriculture can be achieved by implementing management systems on only about 12 percent of cultivated cropland in the watershed, given baseline levels of practice implementation. Nutrient management and erosion control practices, either alone or in combination, are the predominant measures taken to meet program goals. The results also indicate that the Susquehanna basin is where most abatement would ideally come from. Baseline conditions indicate that this region is in the greatest need of improvement, so this finding is not unexpected.

What kind of policy might achieve this outcome? With perfect information, a resource management agency could set discharge limits for each field (performance-based). However, policymakers generally lack complete information on the cost faced by farmers for implementing new practices. With a tax on pollution emissions, policymakers need not know private landowners' costs. We found that tax rates of $2.36 per pound of delivered N, $39.98 per pound of delivered P, and $200.08 per delivered ton of sediment would achieve the TMDL goals at least cost. However, such taxes have never been used in the United States.

Another option would be to allocate pollution allowances equal to agriculture's loading cap (Load Allocation) to fields and let farms trade allowances in a point-nonpoint trading program. Such a cap-and-trade program would achieve the optimal solution, assuming no market "friction," but this would require that agriculture be regulated like point sources. Again, this is not a very likely scenario.

Focusing on Practices May Be More Practical

An alternative to performance-based approaches is to provide incentives for improved management on the basis of field characteristics and implemented practices, as opposed to reductions in pollution achieved (design-based). Theory tells us that such policies are generally less efficient than performance-based policies when transaction costs are ignored, and that is what we found. Simply requiring all cropland to adopt the full suite of conservation measures considered will more than meet the TMDL goals, but at a very high cost (more than 23 times the Optimal scenario).

Efficiency can be improved by targeting practices to cropland that has lower unit abatement costs. This can be accomplished by identifying observable factors or proxies that are closely related to pollutant creation and delivery. Nonpoint-source contribution in a watershed is not evenly distributed but tends to originate from a limited number of locations (Diebel et al., 2008), as indicated by the unequal geographic distribution of treatment under our Optimal scenario. We explored several potential proxies for pollutant delivery in order to target fields with higher pollution losses and lower unit abatement costs. As expected, focusing only on the characteristics of the land, such as erodibility, and ignoring management measures was not cost-effective. Taking into account distance to water and an NRCS assessment of treatment needs provided the least cost design-based approach (of the options we evaluated) for meeting the TMDL.

However, enhanced targeting did not approach the efficiency of the Optimal scenario. Our best enhanced targeting scenario was still about six times more costly than the Optimal scenario. The benefit of performance based policies was demonstrated not only by the lower overall cost, but by the smaller average per-acre treatment costs ($108 per acre vs. $248 per acre). Since performance-based policies induce those fields that can produce the most abatement at least cost to implement management, fewer management systems need to be adopted. Because the relatively simple targeting criteria of the design-based approaches could not target the same cropland receiving management in the Optimal scenario, more treatment was needed per acre and more acres needed to be treated in order to meet the TMDL abatement goals.

Further research with additional data on management options and observable proxies for water quality impacts from fields could improve the recommended targeting criteria. In addition, the development of acceptable modeling tools could have major benefits, enabling the implementation of performance-based approaches. Water quality goals could then be met with fewer acres treated and less extensive treatment on those acres, which is particularly desirable under tight conservation budgets.

Fostering Political Agreement May Require Some Sacrifice in Efficiency

Focusing the bulk of management changes in a particular region may seem inequitable, especially if the region does not enjoy the benefits of improved water quality (as in the case of Pennsylvania). “Spreading the pain” over the entire watershed may be politically desirable to achieve buy-in from all the CBW States. However, as demonstrated with the design of performance-based approaches, a policy becomes increasingly less efficient as it engages fields in parts of the watershed that contribute relatively little to the problem. There can be a significant tradeoff between efficiency (least cost) and equity (treating political jurisdictions more equally). Still, there may be local benefits from improving cropland management over a wider area that we do not account for in our analysis.

Purely Voluntary Approaches Have Some Notable Drawbacks

Meeting the TMDL requires significant reductions in agricultural nonpoint-source pollution. The history of voluntary approaches for meeting watershed water quality goals is not encouraging. Under a voluntary approach, it is difficult or expensive to entice those farmers who can provide the most abatement at least cost to enroll in programs and to adopt the most cost-effective practices for improving water quality. Our models indicate that when farmers are allowed to select which management systems to implement, the TMDL goals are usually not met. Farmers adopted those systems that are least costly to them for meeting the policy, and not necessarily those that reduced pollution at least cost. It may be possible to address this through changes in the way conservation resources are directed. For example, our analysis indicated that the Susquehanna watershed can provide much abatement at a relatively low cost. The proactive development of relationships by resource managers with those farms in the watershed most in need of improvement (rather than waiting for farmers to ask for assistance) and the provision of appropriate education, technical, and financial assistance could greatly improve the efficiency of a voluntary approach. While the administrative costs may be higher, the benefits in terms of increased program efficiency may be significant.

Using Stringent Baselines to Obtain Additional Nonpoint-Source Abatement Hurts Trading

In the absence of regulations on nonpoint-source pollution, adopting stringent eligibility baselines for nonpoint sources in a point-nonpoint trading program in order to obtain additional abatement for meeting a TMDL's Load Allocation could seriously restrict trading while obtaining very little additional abatement. A strict baseline that would require most farms to adopt additional abatement measures in order to trade so restricts credit supply that few credits are traded, and little "extra" abatement is provided for Load Allocation. On the other hand, a less stringent baseline that most farms could meet benefits the credit market and encourages extra abatement that goes to Load Allocation Point sources benefit most when current practices serve as the baseline (no abatement for Load Allocation). An alternative approach could be to use incentives such as traditional conservation programs to encourage abatement for Load Allocation rather than trying to use a trading program to meet multiple goals. Our analysis shows that there is enough potential pollution abatement from nonpoint sources for both meeting the TMDL load allocation and for supporting a trading market with relatively low-cost credits.

Achieving a Better Manure Nutrient Balance

Improving the management of manure is an important part of State plans for meeting the TMDL. The proper implementation of nutrient management practices on animal operations cannot occur until excess manure nutrients are moved to where they can be applied according to standards of a nutrient management plan. This additional cost for complying with a nutrient management plan is an important consideration for animal producers. Such costs range between \$18 million and \$22 million per year in the CBW, assuming half of crop producers are willing to accept manure as a nutrient source. Off-farm hauling costs could be mitigated through education, technical assistance, and financial incentives related to manure management that increase the willingness of crop producers to substitute manure nutrients for commercial fertilizer. Targeting program efforts to areas where competition for suitable land is greatest may achieve the greatest benefit. The cost of such efforts would have to be weighed against the benefits from reduced hauling. Public entities may also help to facilitate market exchanges for manure. A 2011 follow-up survey to the 2003-2006 NRCS CEAP survey found that cropland receiving manure from off-farm sources increased from 17 percent of manured cropland in 2006 to 34 percent in 2011 (USDA, NRCS, 2013). This could be in response to State efforts to encourage more effective use of manure resources under the TMDL.

Developing alternative uses for manure, such as biomass energy, could reduce the hauling problem while providing additional income for animal producers. However, farmers who have come to rely on manure as a source of nutrients and a valuable soil amendment may face rising manure prices or limits on manure supplies as manure is diverted for power generation. Such costs would have to be weighed against the benefits of local energy production. Given the costs of alternative fuels such as natural gas and concerns over local air quality, the current outlook for biomass options is uncertain.

Reductions in manure nutrient production may be part of a broader solution to addressing nutrient imbalances in the watershed. Changes in feed management and other animal production practices have contributed to reduced nutrient loss in the animal waste. Reducing herd size is a more drastic approach for achieving a manure-nutrient balance, but there is precedent in Florida. However, the savings in reduced hauling costs would be outweighed by the value of lost production.

REFERENCES

Aillery, M., N. Gollehon, and V. Breneman. 2005. *Technical Documentation of the Chesapeake Bay Regional Model: Model Version – ERS-RED Manure Management Project*. TB-1923, U.S. Department of Agriculture, Economic Research Service.

Aillery, M., N. Gollehon, V. Breneman, and S. Bucholtz. 2009. "Modeling Firm Spatial Interdependence using National Data Coverages: A Regional Application to Manure Management," *Natural Resources Modeling* 22(1): 42-66.

Aillery, M., N. Gollehon, R. Johansson, J. Kaplan, N. Key, and M. Ribaudo. 2005. *Managing Manure to Improve Air and Water Quality*. ERR-9, Economic Research Service, U.S. Department of Agriculture.

American Farmland Trust (AFT). 2013. *Involving Agriculture in Water Quality Trading Markets*. Center for Agriculture in the Environment, DeKalb, IL.

Arnold, J.G., R. Srinivasan, R.S. Muttiah, and P.M. Allen. 1999. "Continental scale simulation of the hydrologic balance," *Journal of the American Water Resources Association* 35(5): 1037-1052.

Arnold, J.G., and N. Fohrer. 2005. "SWAT2000: current capabilities and research opportunities in applied watershed modeling," *Hydrological Processes* 19(3): 563-572.

Babcock, B. A. P. Lakshminarayan, J. Wu, and D. Zilberman. 1997. "Targeting Tools for the Purchase of Environmental Amenities," *Land Economics* 73 (3): 325-339.

Baker, M.E., D.E. Weller, and T.E. Jordan. 2007. "Effects of stream map resolution on measures of riparian buffer distribution and nutrient retention potential," *Landscape Ecology* 22: 973-992.

Bosch, D.J., J.W. Pease, S.S. Batie, and V.O. Shanholtz. 1992. *Crop Selection, Tillage Practices, and Chemical and Nutrient Applications in Two Regions of the Chesapeake Bay Watershed.* Bulletin 176. Virginia Water Resources Research Center, Virginia Polytechnic Institute and State University, Blacksburg, VA.

Braden, J.B., and K. Segerson. 1993. "Information Problems in the Design of Nonpoint-Source Pollution Policy," in *Theory, Modeling, and Experience in the Management of Nonpoint-Source Pollution.* C.S. Russell and J.F. Shogren (eds.), Boston: Kluwer Academic Publishers.

Braden, J. B., A. Bozaher, G. Johnson, and D. Miltz. 1989. "Optimal Spatial Management of Agricultural Pollution," *American Journal of Agricultural Economics* 71: 404-413.

Camboni, S. M., and T.L. Napier. 1994. *Socioeconomic and Farm Structure Factors Affecting Frequency of Use of Tillage Systems.* Invited paper presented at the Agrarian Prospects III symposium, Prague, Czech Republic.

Catma, Serkan, and Alan R. Collins. 2011. "Phosphorus Imbalances in the Chesapeake Bay Watershed: Can Forestland and Manure Processing Facilities Be the Answers?" *Agricultural and Resource Economics Review* 40 (1):1-10. April.

Chesapeake Bay Program. 2011. *Bay Barometer: A Health and Restoration Assessment of the Chesapeake Bay and Watershed in 2010*. Annapolis, MD.

Chesapeake Bay Program. 1999. *The State of the Chesapeake Bay*. CBP/TRS 222/108. Annapolis, MD.

CBP (Chesapeake Bay Program). 1983. *1983 Chesapeake Bay Agreement*. Annapolis, MD.

CBP. 1987. *1987 Chesapeake Bay Agreement*. Annapolis, MD.

CBP. 1992. *Chesapeake Bay Agreement: 1992 Amendments*. Annapolis, MD.

CBP. 2000. *2000 Chesapeake Bay Agreement*. Annapolis, MD.

CBP. 2004. *Backgrounder: What Are Tributary Strategies?* Annapolis, MD.

CBP. 2007. *Chesapeake Bay 2006 Health and Restoration Assessment*. Annapolis, MD.

Claassen, R., and R.D. Horan. 2001. "Uniform and Non-uniform Second-Best Input Taxes: The Significance of Market Price Effects on Efficiency and Equity," *Environment and Resource Economics* 19 (1): 1-22.

Code of Federal Regulations. 1996. 7 C.F.R. Subpart B-Highly Erodible Land Conservation.

Diaz-Zorita, M., J.H. Grove, L. Murdock, J. Herbeck, and E. Perfect. 2004. "Soil structure disturbance effects on crop yields and soil properties in a no-till production system," *Agronomic Journal* 96:1651-1659.

Diebel, M.W., J.T. Maxted, P.J. Nowak, and M.J. Vander Zanden. 2008. "Landscape Planning for Agricultural Nonpoint Source Pollution Reduction I: A Geographical Allocation Framework," *Environmental Management* 42:789-802.

Fleming, R.A., and R.M. Adams. 1997. "The Importance of Site-Specific Information in the Design of Policies to Control Pollution," *Journal of Environmental Economics and Management* 33: 347-358.

Gassman, P.W., M.R. Reyes, C.H. Green, and J.G. Arnold. 2007. "The Soil and Water Assessment Tool: Historical development, applications and future research directions," *Transactions of the American Society of Agricultural and Biological Engineers* 50(4): 1211-1250.

Gassman, P.W., J.R. Williams, S. Wang, A. Saleh, E. Osei, L. Hauck, C. Izaurralde, and J. Flowers. 2009. *The Agricultural Policy Environmental Extender (APEX) model: An emerging tool for landscape and watershed environmental analyses*. Technical Report 09-TR 49. CARD, Iowa State Univ., Ames, IA.

Genskow, K.D. 2012. "Taking stock of voluntary nutrient management: Measuring and tracking change," *Journal of Soil and Water Conservation* 67(1):51-58.

Ghosh, G., M. Ribaudo, and J. Shortle. 2011. "Baseline Requirements can hinder trades in water quality trading programs: Evidence from the Conestoga watershed," *Journal of Environmental Management* 92 (8): 2076-2084.

Gillenwater, M. 2012. "What is Additionality: Part 2: A framework for more precise definitions and standardized approaches." Discussion Paper No. 002. GHG Management Institute, Silver Spring, MD.

Gollehon, N., M. Caswell, M. Ribaudo, R. Kellogg, C. Lander, and D. Letson. 2001. *Confined Animal Production and Manure Nutrients.* AIB 771, U.S. Department of Agriculture, Economic Research Service.

Hoag, D.L., and J.S. Hughes-Popp. 1997. "Theory and Practice of Pollution Credit Trading in Water Quality Management," *Review of Agricultural Economics* 19 (2): 252-262.

Horan, R.D., and J.S. Shortle. 2011. "Economic and ecological rules for water quality trading," *Journal of the American Water Resources Association* 47 (1): 59-69.

Jackson-Smith, D.B., M. Halling, E. de la Hoz, J.P. McEvoy, and J.S. Horsburgh. 2010. "Measuring conservation program best management practice implementation and maintenance at the watershed scale," *Journal of Soil and Water Conservation* 65 (6): 413-423.

Kellogg, R.L., D.C. Moffitt, and N. Gollehon. Forthcoming. *Estimates of Recoverable and Non-Recoverable Manure Nutrients Based on the Census of Agriculture*. Natural Resources Conservation Service, U.S. Department of Agriculture.

Kellogg, R.L., C.H. Lander, D.H. Moffitt, and N. Gollehon. 2000. "Manure nutrients relative to the capacity of cropland and pastureland to assimilate nutrients—Spatial and temporal trends for the United States." Natural Resources Conservation Service, U.S. Department of Agriculture.

King, D.M. 2005. "Crunch Time for Water Quality Trading," *Choices* 20: 71-76.

King, D.M., and P.J. Kuch. 2003. "Will Nutrient Credit Trading Ever Work? An Assessment of Supply and Demand Problems and Institutional Obstacles," *Environmental Law Reporter* 33: 10352-10368.

Kling, K. 2011. "Economic Incentives To Improve Water Quality in Agricultural Landscapes: Some New Variations on Old Ideas," *American Journal of Agricultural Economics* 93 (2): 297-309.

Kramer, J.M. 2003. *Lessons from the Trading Pilots: Applications for Wisconsin Water Quality Trading Policy*. Fox-Wolf Watershed Alliance, Appleton, WI.

Lankowski, J., and A. Cattaneo. 2010. *Guidelines for Cost-effective Agri-Environmental Policy Measures*. Organization for Economic Cooperation and Development, Paris.

Lichtenberg, E., D. Parker, and S. Lane. 2010. *Best Management Practice Use and Nutrient Management in Maryland: A 2010 Snapshot.* Center for Agricultural and Natural Resource Policy, University of Maryland.

Lynne, G.D., J.S. Shonkwiler, and L.R. Rola. 1988. "Attitudes and farmer conservation behavior," *American Journal of Agricultural Economics* 70: 12-19.

Marshall, E., and M. Weinberg. 2012. *Baselines in Environmental Markets: Tradeoffs Between Cost and Additionality*. Economic Brief 18, Economic Research Service, U.S. Department of Agriculture.

Napier, T. L., and S.M. Camboni. 1993. "Use of Conventional and Conservation Practices among Farmers in the Scioto River Basin of Ohio," *Journal of Soil and Water Conservation* 48: 432-438.

National Research Council (NRC). 2011. *Achieving Nutrient and Sediment Reduction goals in the Chesapeake Bay: An Evaluation of Program Strategies and Implementation*. National Academy of Sciences, Washington, DC.

Nowak, P.J. 1987. "The Adoption of Agricultural Conservation Technologies: Economic and Diffusion Explanations," *Rural Sociology* 52: 208-220.

Oates, W. 1995. "Green Taxes: Can We Protect the Environment and Improve the Tax System at the Same Time?" *Southern Economic Journal* 61: 915-922.

Osmond, D., D. Meals, D. Hoag, M. Arabi, A. Luloff, G. Jennings, M. McFarland, J. Spooner, A. Sharpley, and D. Line. 2012. "Improving conservation practices programming to protect water quality in agricultural watersheds: Lessons learned from the National Institute of Food and Agriculture-Conservation Effects Assessment Project," *Journal of Soil and Water Conservation* 67 (5): 122A-127A.

Pease, J.J., Arogo Ogejo, R. Maguire, D. Bosch, J. Farris, and J. O'Neill. 2012. *Evaluating Net Benefits/Impacts of a Shenandoah Valley Poultry Litter to Energy Power Plant*. Report to Virginia Dept. of Environmental Quality. April.

Pendell, D.L., J.R. Williams, C.W. Rice, R.G. Nelson, and S.B. Boyles. 2006. "Economic feasibility of no-tillage and manure for soil carbon sequestration in corn production in northeastern Kansas," *Journal of Environmental Quality* 35: 1364-1373.

Purvis, A., W.G. Boggess, C.B. Moss, and J. Holt. 1995. "Technology Adoption Decisions Under Irreversibility and Uncertainty: an Ex Ante Approach," *American Journal of Agricultural Economics* 77 (3): 541-551.

Ribaudo, M.O. 1986. "Consideration of Offsite Impacts in Targeting Soil Conservation Programs," *Land Economics* 62 (4): 402-411.

Ribaudo, M.O. 1989. "Targeting the Conservation Reserve Program To Maximize Water Quality Benefits," *Land Economics* 65 (4): 320-332.

Ribaudo, M. O., and R.D. Horan. 1999. The Role of Education in Nonpoint Source Pollution Control Policy. *Review of Agricultural Economics* 31 (2), 331-343.

Ribaudo, M., J. Delgado, L. Hansen, M. Livingston, R. Mosheim, and J. Williamson. 2011. *Nitrogen in Agricultural Systems: Implications for Conservation Policy*. Economic Research Report ERR-127, Economic Research Service, U.S. Department of Agriculture.

Ribaudo, M., N. Gollehon, M. Aillery, J. Kaplan, R. Johansson, J. Agapoff, L. Christensen, V. Breneman, and M. Peters. 2003. *Manure Management for Water Quality: Cost to Animal Feeding Operations of Applying Manure Nutrients to Land*. Agricultural Economics Report, AER-824, Economic Research Service, U.S. Department of Agriculture.

Ribaudo, M.O., and J. Gottlieb. 2011. "Point-Nonpoint Trading – Can it Work?" *Journal of the American Water Resources Association* 47 (1): 5-14.

Ribaudo, M. O., R.D. Horan, and M.E. Smith. 1999. *Economics of Water Quality Protection from Non-point Sources: Theory and Practice*. AER-782, U.S. Department of Agriculture, Economic Research Service.

Schmitz, A., W.G. Boggess, and K. Tefertiller. 1995. "Regulations: Evidence from the Florida Dairy Industry," *American Journal of Agricultural Economics* 77 (5): 1166-1171.

Schwartz, S.S. 2010. "Optimization and Decision Heuristics for Chesapeake Bay Nutrient Reduction Strategies," *Environmental Modeling and Assessment* 15 (5): 345-359.

Seaber, P.R., F.P. Kapinos, and G.L. Knapp. 1987. "Hydrologic Unit Maps." Water-Supply Paper 2294, U.S. Geological Survey: Reston, VA.

Segerson, K., and D. Walker. 2002. "Nutrient Pollution: An Economic Perspective," *Estuaries* 25 (4): 797-808.

Shortle, J.S. 2013. "Economics and environmental markets: Lessons from water quality trading," *Agricultural and Resource Economics Review* 42 (1): 57-74.

Shortle, J., and R. Horan. 2001. "The Economics of Nonpoint Pollution," *Journal of Economic Surveys* 15: 255-290.

Shortle, J.S., and D.G. Abler. 1997. "Nonpoint Pollution," in *International Yearbook of Environmental and Natural Resource Economics*. H. Folmer and T. Teitenberg (editors), Cheltenham, UK: Edward Elgar.

Shortle, J., M. Ribaudo, R. Horan, and D. Blandford. 2012. "Reforming Agricultural Nonpoint Pollution Policy in an Increasingly Budget-Constrained Environment," *Environmental Science and Technology* 46 (3): 1316-1325.

Sneeringer, S. 2013. "Differences between Livestock and Crop Producers' Participation in Nutrient Trading," *Applied Economic Perspectives and Policy* 35(2): 296-321.

Srinivasan, R.S., J.G. Arnold, and C.A. Jones. 1998. "Hydrologic modeling of the United States with the Soil and Water Assessment Tool," *International Journal of Water Resources Development* 14(3): 315-325.

Stephenson, K., S. Aultman, T. Metcalfe, and A. Miller. 2010. "An evaluation of nutrient nonpoint offset trading in Virginia: A role for agricultural nonpoint sources?" *Water Resources Research* 46: 1-11.

U.S. Department of Agriculture. 2006. "Partnership Agreement Between the United States Department of Agriculture Natural Resources Conservation Service and the United States Environmental Protection Agency Office of Water." http://www.epa.gov/owow/watershed/ trading/mou061013.pdf.

U.S. Department of Agriculture, Natural Resources Conservation Service. 2011. *Assessment of the Effects of Conservation Practices on Cultivated Cropland in the Chesapeake Bay Region*. Conservation Effects Assessment Project.

U.S. Department of Agriculture, Natural Resources Conservation Service. 2013. *Impacts of Conservation Adoption on Cultivated Cropland in the Chesapeake Bay Region, 2003-06 to 2011*. Conservation Effects Assessment Project.

U.S. Department of Agriculture, Natural Resources Conservation Service. 2012. *Chesapeake Bay 2011 Activities Report*. http://www.nrcs.usda.gov/ Internet/FSE_DOCUMENTS/stelprdb1047307. pdf.

U.S. Environmental Protection Agency. 2011. *An Optimization Approach to Evaluate the Role of Ecosystem Services in Chesapeake Bay Restoration Strategies*. EPA/600/R-11/001, Office of Research and Development.

U.S. Environmental Protection Agency. 1983. *Chesapeake Bay: A Framework for Action.* Chesapeake Bay Program Office: Annapolis, MD.

U.S. Environmental Protection Agency. 2007. *Water Quality Trading Toolkit for Permit Writers*. EPA 833-R-07-004; Office of Water.

U.S. Environmental Protection Agency. 2008. *EPA Water Quality Trading Evaluation: Final Report.*

U.S. Environmental Protection Agency. 2010a. *Chesapeake Bay Compliance and Enforcement Strategy*. Office of Enforcement and Compliance Assurance.

U.S. Environmental Protection Agency. 2010b. *Chesapeake Bay Total Maximum Daily Load for Nitrogen, Phosphorus, and Sediment.* http://www.epa.gov/reg3wapd/tmdl/ChesapeakeBay/ tmdlexec.html.

U.S. Environmental Protection Agency, Office of Inspector General; U.S. Department of Agriculture, Office of Inspector General. 2006. *Inspector General Evaluation Report: Saving the Chesapeake Bay Watershed Requires Better Coordination of Environmental and Agricultural Resources*; EPA OIG Report No. 2007-P-00004, USDA OIG Report No. 50601-10-IIq.

Veith, T.L., M.L. Wolfe, and C.D. Heatwole. 2004. "Cost-Effective BMP Placement: Optimization Versus Targeting," *Transactions of the American Society of Agricultural Engineers* 47 (5): 1585-1594.

Wainger, L. 2012. "Opportunities for Reducing Total Maximum Daily Load (TMDL) Compliance Costs: Lessons from the Chesapeake Bay," *Environmental Science & Technology* 46: 9256-9265.

Weaver, R.D. 1996. "Prosocial Behavior: Private Contributions to Agriculture's Impact on the Environment," *Land Economics* 72 (2): 231-247.

Weersink, A., J. Livenois, J. Shogren, and J. Shortle. 1998. "Economic Incentives for Pollution Control in Agriculture," *Canadian Public Policy* 24 (3): 309-327.

White House. 2009. Chesapeake Bay Protection and Restoration. Executive Order 13508. http://www.whitehouse.gov/the_press_office/ Executive-Order-Chesapeake-Bay-Protection-and-Restoration/.

Williams, J.R., R.C. Izaurralde, and E.M. Steglich. 2008. *Agricultural Policy/Environmental eXtender Model: Theoretical documentation version 0604*. BREC Report # 2008-17. Temple, TX: Texas AgriLIFE Research, Texas A&M University, Blackland Research and Extension Center.

Woodward, R.T., and R.A. Kaiser. 2002. "Market Structures for U.S. Water Quality Trading," *Review of Agricultural Economics* 24: 366-383.

APPENDIX 1. MODELING SCENARIOS FOR IMPROVING CROPLAND NUTRIENT MANAGEMENT

The representative farm is the primary modeling unit. Expansion factors provided by NRCS Conservation Effects Assessment Program (CEAP) are used to describe all agricultural land within each four-digit Hydrologic Unit Code (HUC). In this way, the observations account for heterogeneity in nutrient losses and adoption costs for best management systems for all agricultural land within the Bay watershed. We restricted our analysis to major field crops, which constituted 97.9 percent of agricultural land in the Bay. We excluded fields planted to vegetables, as we have less confidence that our cost data are representative of these specialized crops.

In order to meet the Bay TMDL targets, each representative farm can either stay in the baseline or adopt one or more of the three practices: nutrient management, cover crops, and soil erosion controls. The adoption costs and pollution reduction benefits generated by practice adoption are unique to each representative farm and depend on the physical conditions of the land and the current practices in place. We assume in our analysis that farms do not enter or exit the sector in response to the policy.

Design-Based Scenarios

The benefits and costs of the design-based scenarios were calculated by selecting the observations meeting the relevant conditions, then adding up the total costs and pollution reduction benefits.

Performance-Based Scenarios

The performance-based scenarios, other than the tax, were solved using mixed integer programming. The tax model was solved using a two-stage process that is explained in more detail below in the description of that model.

Optimal

The optimization model is designed to minimize the total cost of reducing nutrient and sediment losses to the watershed, subject to meeting the Bay TMDL pollution reduction targets for N, P, and sediment of 25, 24, and 20 percent. Let $r^k, k \in \{N,P,Sediment\}$ represent the pollutant emission goals for the TMDL. That is, $\{r^N, r^P, r^{Sediment}\} = \{0.75, 0.76, 0.80\}$. Each representative farm can choose one of eight BMP sets: the baseline (no change); nutrient management; cover crops; soil erosion control practices; and each of the four possible combinations of those practices. The model allocates BMPs across farms to minimize the objective function expression:

$$\min_x \sum_i \sum_j c_{ij} x_{ij}$$

where C_{ij} is the annual adoption cost associated with farm $i \in \{1,...,M\}$ and BMP set $j \in \{0,...,7\}$ and x_{ij} is a binary variable equal to one if practice set j is adopted. Costs reflect the total costs to the farmer of adopting the practice set, including practice implementation costs, the opportunity costs of changes in nutrient applications and crop yields, and costs associated with increased risk.

Associated with each practice adoption choice x_{ij} is a set of field pollutant emissions delivered to tidal portions of the Chesapeake Bay d_{ij}^k, where $k = N,P,Sediment$. Let d_{i0}^k represent delivered emissions for pollutant k by observation i using baseline practices and let the aggregate level of emissions for pollutant k be given by $e^k = \Sigma_i d_{i0}^k$

We have three nutrient reduction constraints to ensure that the solution meets the goals of the TMDL:

$$\sum_i \sum_j d_{ij}^k x_{ij} \le r^k base^k$$

We have 756 farm practice constraints to make sure the model picks exactly one practice per observation:

$$\Sigma_i x_{ij} = 1.$$

Regional

Let h index the four-digit HUCs in the watershed and let HUG be a binary variable indicating the HUC in which the representative farm operates. The objective function and the farm practice constraints are the same as in the optimal scenario, but the regional model requires twelve (h x k) nutrient reduction constraints:

$$\sum_i \sum_j HUC_h d_{ij}^k x_{ij} \leq r^k base_h^k$$

where

$$base_h^k = \Sigma_i HUC_h d_{i0}^k \, .$$

Farm Level

Minimizing the costs of meeting the pollutant reduction goals at the farm level is complicated by the fact that it is not possible for all farms to achieve the reductions for any or all of the pollutants. We relax this requirement so that farms must adopt new suites of practices (adopt practice suite $j \neq 0$) if its adoption satisfies at least one of the pollution reduction goals. Let be a variable that effectively acts as an extra cost of remaining in the baseline for each pollutant. Let be a scalar that is sufficiently large to make the penalty outweigh the costs of adopting practices. We used k = 1,000,000,000,000 to estimate the model. The objection function was specified as:

$$\min_{x,\, penalty} \sum_i \sum_j c_{ij} x_{ij} + k \sum_i \sum_k \frac{penalty_i^k}{base^k}$$

For consistency, the $penalty_i^k$ variables are also added to the (i x k) nutrient reduction constraints:

$$\Sigma_j d_{ij}^k \leq r^k d_{i0}^k + penalty_i^k$$

Tax

The optimization model for the pollutant tax scenario is designed to produce a solution similar to the solution produced by the optimal scenario. A two-stage model was used to solve for the optimal tax rates and to solve for the optimal distribution of practice adoption under these tax rates. To the cost-minimization objective, we add tax rates on each unit of pollutant delivered to the tidal waters by observation i, e_i^k:

$$\min_{x,e} \sum_i \sum_j c_{ij} x_{ij} + \sum_i \sum_k t^k e_i^k$$

We use $i \times k$ emissions constraints to set the emissions associated with practice suite $j, d_{ij}^k x_{ij}$, equal to emissions e_i^k:

$$\sum_j d_{ij}^k x_{ij} = e_i^k$$

Similar to other performance-based models, we have i farm practice constraints so that each observation implements exactly one suite of practices j:

$$\sum_j x_{ij} = 1$$

In the first stage, we also have k emissions reduction constraints so that aggregate pollutant emissions to tidal waters meet the TMDL goals:

$$\Sigma_i e_i^k \leq r^k base^k$$

In order to address the complexity added by the addition of emissions to the objective, a feasible solution to the first-stage model is found using relaxed nonlinear mixed integer programming. The optimal tax rates are set equal to the marginal values of the emissions constraints in the first stage. In the second stage, the emissions reduction constraints are dropped from the model and the optimal distribution of practice adoption is found using mixed integer programming.

Model Data

The Chesapeake Bay watershed models rely primarily on data from the CEAP assessment of the effects of conservation practices on agricultural land in the watershed. For the analysis, 756 National Resources Inventory (NRI) sampling points in the watershed were treated as representative farms growing major field crops (excluding vegetables) using sampling weights to describe farm production and environmental conditions on 97.9 percent of agricultural land in the watershed. Cost data and other information reflecting conditions in the watershed States (DE, MD, NY, PA, VA, WV) were obtained from various sources, including State Watershed Implementation Plans, contract data from USDA's Environmental Quality Incentives Program (EQIP), the Chesapeake Bay Program, published literature, and subject matter specialists within government and various universities.

CEAP Model Data

Data from the CEAP model were used to assess the effects of practice adoption on pollutants delivered to the tidal waters of the basin. We use HUC-level delivery ratios to scale edge-of-field pollutant losses to the tidal waters of the Bay. The CEAP model data also provide acreage expansion factors to scale representative farms to the HUC-level aggregates. Edge-of-field nitrogen losses are given by the total N lost in surface pathways and in subsurface water flow. Edge-of-field phosphorus losses are given by P lost with sediment and total soluble P loss. Edge-of-field sediment losses are given by the Revised Universal Soil Loss Equation (RUSLE) erosion rate for the cropped region.

Production Cost Data

We considered only the costs of adopting additional practices: cover crops, soil erosion controls, and nutrient management. For each practice, we account for changes in fertilizer application rates, changes in crop yields, annualized practice installation costs, and costs related to uncertainty and risk. For soil erosion controls, which take arable land out of production, we also consider opportunity costs of the land. Policy transaction costs are not included.

For each of the eight scenarios, we use NRCS-calculated fertilizer application rates, crop yield per acre, and the proportion of land cropped. For each year in the crop rotation, we calculated the value of output and fertilizer input costs. These data were averaged to represent production costs as the average annual cost of implementing new practices. The data from CEAP contain 28 unique crop rotations. We estimated the crop value based on reported 2010 State-level crop prices provided by USDA's National Agricultural Statistics Service (NASS). For some crops, State-level prices were not available, so we represent the values of these crops using the U.S. price.

Fertilizer costs are based on reported 2010 prices by NASS. Nitrogen price reflects the U.S. average price ($499 per ton) for anhydrous ammonia, which is 82 percent N, or a price per active ingredient of $0.30 per pound of N. Phosphorus price reflects the price per ton of superphosphate (46 percent P), or an active ingredient price of $0.55 per pound of P. From CEAP, we have information on whether manure was used, but we have incomplete data on animal production by the land operators and do not know whether the manure applied was produced on the operation. Without information on the price paid for the manure applied, we assume the cost of nutrients from manure equals the cost of nutrients from chemical fertilizers.

We account for the opportunity cost of land taken out of production for soil erosion controls based on 2010 State-level land rental rates for irrigated and non-irrigated land reported by NASS. We obtained annualized, State-level installation costs for nutrient management and cover crops from the World Resources Institute (WRI).

The specific practices implemented as soil erosion controls in the CEAP report depend on the slope, inherent vulnerability of the soil, and proximity to water. Soil erosion controls include grass buffers, contouring, and terracing. We obtained annualized, State-level installation costs for grass buffers from WRI. Terracing costs were obtained from EQIP contract data and State payment rates. For contouring, we relied on the formulas given in Appendix D of NRCS (2003)[17] and State EQIP payment rates.

Spatial Land Data

Cost data were scaled to the HUC level based on the distribution of cropland within each State and HUC boundary. To assess the spatial pattern of cropland within each State and HUC boundary, we use the 2007 U.S. Geological Service's National Land Cover Dataset. This dataset is based primarily on 2001 Landsat thematic mapper imagery at 30-meter resolution, classified in 16 land-use categories for land within the study region. The U.S. Geological Survey Digital Line Graph state boundaries and the 8-digit USGS hydrologic units of the Chesapeake Bay Basin can be found at http://www.chesapeakebay.net/data.

APPENDIX 2—UPDATES TO THE CHESAPEAKE BAY REGIONAL MANURE TRANSPORT MODEL

The Chesapeake Bay Regional Manure Transport Model, developed for earlier ERS analysis of manure-nutrient management (Ribaudo et al., 2003; Aillery et al., 2005), was updated and extended for use in the Chesapeake Bay TMDL study. Documentation of the basic model structure and supporting data is available in published form (Aillery et. al., 2005, Aillery et al., 2009). Here we highlight several changes in the model data and modeling framework that are specific to the TMDL analysis presented in this report.

Model Data

NRCS animal manure nutrients. County-level data from the 2007 Census of Agriculture processed by NRCS (hereafter referred to as the NRCS dataset) provide the data foundation for the ERS regional manure model. The agricultural census data were supplemented with other biophysical and production information on the U.S. livestock sector. The NRCS dataset includes estimates of AFO and non-AFO farms, confined and non-confined animal production, recoverable and non-recoverable manure production, nutrient content of manure, cropland assimilative capacity for manure nutrients on AFO and non-AFO farms, and excess manure nutrients produced on AFO farms. Estimates were derived from individual farm-level survey data and aggregated by farm type to counties and 6-digit hydrologic sub-basins for reporting and research purposes. Data estimation procedures and assumptions are detailed in Kellogg et al. (2014).

Previous applications of the regional manure model drew on an earlier version of the NRCS database, developed from the 1997 census (Kellogg et al., 2000). The 2007 data values reflect the changing structure of the animal sector and evolving technology use in animal production and manure handling. In some instances, NRCS estimation procedures were modified to provide improved measures—as in the case of nutrient assimilative capacity on animal feeding operations (AFOs) and resulting manure-nutrient excess. Census survey data on farmland receiving manure, included separately for AFOs and non-AFOs in the NRCS database, provide the basis for 2007 base conditions under the TMDL analysis.

ERS analysis was further informed by NRCS assumptions developed for a national simulation analysis of county-level excess manure nitrogen (Kellogg et al., 2014). Key assumptions involved nitrogen application-uptake ratios used in calculating applied manure (with and without full conservation treatment), the share of AFO cropland acreage considered to be soil-P limiting or highly erodible in the basin (from the Chesapeake Bay CEAP study), the share of permanent pastureland assumed unsuitable for manure on non-AFO farms, and the willingness of landowners to accept manure. In some local areas, the application of biosolids from wastewater treatment plants competes for available agricultural land, although we do not account for this in our study. In other areas, animal manure has been used on commercial forestland (Catma and Collins, 2011), which is not considered part of the potential land base assumed in our study. To the extent possible, the ERS analysis was designed to maintain consistency with NRCS data and simulation assumptions.

GIS Area-to-Distance Functions. Land competition for manure spreading reflects the spatial relationship among AFOs with excess manure, and the extent and pattern of available land for manure spreading. The GIS-derived area-to-distance component—an innovative feature of the ERS regional manure model—was designed to capture the competition for land, which is an important determinant of manure hauling distance and cost.

Area-to-distance relationships were updated for the TMDL analysis to more accurately reflect current conditions in the regional animal sector. The number of confined animal feeding operations, obtained through the NRCS animal-manure database, was derived from reported animal production in the 2007 Census of Agriculture. The spatial land cover of cropland and permanent pasture was drawn from the 2012 Cropland Data Layer (CDL), developed by USDA's National Agricultural Statistics Service (NASS).

Revised estimation procedures for GIS area-to-distance functions provided a more realistic assessment of potential hauling distances for use in the TMDL study. Estimation procedures were improved in several ways:

- The assignment of AFOs by county and sub-basin is potentially more spatially accurate relative to earlier studies based on AFO counts by county, as the 2007 Census of Agriculture collected data to identify farm locations, including AFOs by county and HUC-6.

- Consistent with NRCS assumptions, half of permanent pasture acreage—randomly drawn from pastureland coverage in the watershed—was assumed unsuitable for manure spreading due to terrain, tree cover, soil erodibility, and other factors.
- For each unique intercounty manure transfer from source-county grid to farmland perimeter in receiving counties, distances were calculated based on actual road mileage—in contrast with the prior estimation, which adjusted linear distances by rough factor adjustments to reflect water bodies and other natural barriers.
- For each intercounty transfer, a unique area-to-distance function is estimated for the receiving county, measured from the perimeter point of access to farmland.
- For all intra- and intercounty area-to-distance functions, county-specific circuity factors that convert from linear to road distance, and are estimated from actual road miles, replace State-level circuity factors applied in the prior analyses.

As with the prior analyses, slope coefficients for the area-to-distance functions were adjusted in the simulation model to reflect alternative levels of willingness to accept manure. Lower (higher) WTAM levels result in lesser (greater) available acreage on which to spread manure, as well as longer (shorter) hauls to access a given land area. In addition, the TMDL analysis introduces an index adjustment to intracounty slope coefficients to better reflect AFO size considerations and resulting intensity of onfarm manure production on competition for land. The effect is to slightly increase hauling distances required to access a given land area in counties where average dry-manure tonnage per AFO is greatest.

Other Model Data. Various production and cost data were used in updating the regional manure model for use in the TMDL study. The distribution of manure system types by animal species was revised based on USDA's Agricultural Resource Management Survey (ARMS) data for poultry, hogs, and dairy. Annual returns above operating costs were drawn from livestock sector data in the ERS-REAP national agricultural sector model, based on ERS cost/return estimates and other source; costs per live-animal weight were converted to animal-unit equivalents using conversion factors reported in Kellogg et al. (2014). Chemical fertilizer prices for (elemental) nitrogen and phosphorus, as well as prices paid for poultry litter, were obtained from Pease et al. (2012).

Manure used in non-cropland applications was estimated from available sources, as published estimates are generally unavailable. Poultry litter tonnage used annually for manure pelletization was drawn from the Maryland Watershed Implementation Plan. Manure diversions for small-scale composting were derived from the ARMS poultry survey. The Chesapeake Bay model provided more spatially detailed estimates on annual nitrogen deposition rates by county, which were used in adjusting agronomic rates for applied manure.

Per-unit hauling and application costs in the model largely draw on values compiled for the prior model version. Hauling costs are specified per ton and unit-distance, by manure handling system and distance interval traveled. Land application costs are specified per acre, with and without incorporation/injection. We assume that 50 percent of manure applied to cropland is incorporated under the *BASE2007* scenario (NRCS-CEAP, 2011), increasing to 100 percent under NMP compliance scenarios.

Model Structure

While the basic structure of the ERS regional manure model is essentially the same as that used in previous studies, a number of changes were incorporated for use in the TMDL analysis. Modifications in model structure reflect both the availability of source model data as well as revised objectives of the study analysis. Key model extensions are as follows:

- The model was restructured around four AFO size classes—Large AFO-CAFOs, Medium AFO-CAFOs, Small AFOs, and Very Small AFOs, as defined in the NRCS animal-nutrient data and patterned after EPA regulatory distinctions by farm class. This enables analysis of scenarios differentiated by AFO type, although AFO size distinctions were not a focus of the TMDL analysis.
- The model was extended to consider multiple nonfarm uses of manure, including pelletization, composting, and biomass energy.
- Model extensions also support analysis of animal number reductions as a means of addressing manure excess. This includes estimation of weighted manure-excess per animal-unit and returns per animal-unit forgone.

- Selected output variables are reported at both the county level and 6-digit hydrologic sub-basin, providing greater opportunity to evaluate variation in production conditions and policy impacts across areas of the watershed.

Whereas the 1997 NRCS dataset included estimates of farm-level excess and land assimilative capacity for manure-nutrients under both a nitrogen and phosphorus-based standard, the 2007 NRCS dataset assumes a nitrogen standard. While this limited our capacity in the TMDL study to examine manure hauling implications under a P-based standard, assumed reductions in AFO spreadable land acreage due to P-limited soils addresses the phosphorus concern to some degree.

End Notes

[1] Hydrologic Unit Code (HUC) is a classification system used by the U.S. Geological Survey to catalog watersheds. A 4-digit HUC, of which there are 221 in the United States, is a subregion that includes the area drained by a river system, a reach of a river and its tributaries, a closed basin, or a group of streams forming a coastal drainage area (Seaber et al., 1987). The Chesapeake Bay watershed is made up of four 4-digit HUCs: Susquehanna (0205), Potomac (0206), Upper Chesapeake (0207), and Lower Chesapeake (0208).

[2] We refer to Best Management Systems instead of the more common Best Management Practices (BMP). The management improvements in our analysis consist of suites of practices rather than a single practice.

[3] An alternative approach for estimating the nutrient loadings would have been to apply the Chesapeake Bay watershed model used by EPA to track changes in loadings within the Chesapeake Bay watershed and to track the progress States are making in implementing their WIPs. We chose to use the NRCS modeling system to estimate changes in loadings because it provides field-level output (the EPA model is a county-level model) and thus is better suited to make full use of the richness of the field-level data provided by the CEAP survey. This approach was critical for providing information on the cost-effectiveness of field-level targeting for practices.

[4] Tax rates are the shadow prices for the constraints on nitrogen, phosphorus, and sediment in the optimization model. In simple terms, it is the benefit of relaxing the constraint by one unit.

[5] While we do not have current estimates of AFO assimilative capacity for phosphorus, previous studies suggest that the percent of manure P that could be assimilated onfarm is considerably lower than that for manure N (Ribaudo et al., 2003).

[6] Phytase is a supplemental enzyme added to feedgrain regimes for poultry and hogs to reduce the phosphorus content of manure. Under a P-based nutrient standard, lower manure-phosphorus concentrations allow higher rates of applied manure to cropland. Phytase use has expanded in the Chesapeake Bay watershed in recent years. Analysis of phytase treatment for manure management was examined in prior ERS research (Ribaudo et al., 2003).

[7] Alum is a chemical added to poultry litter primarily to reduce ammonia emissions from poultry houses. It also reduces P runoff from fields receiving litter. Analysis of alum treatment for manure management was examined in prior ERS research (Aillery et al., 2005).

[8] We assume AFOs are randomly distributed on crop and pasture land within the county and HUC-6 sub-basin areas in which they are assigned in the source database (USDA, NRCS, 2012).

[9] Current non-cropland manure uses include approximately 30,000 tons used annually for poultry-litter pelletizing (Perdue Agri-Cycle) in southern Maryland and Delaware. In addition, we assume that small-scale composting accounts for 2 percent of dry poultry litter and 1 percent of other 'semi-dry' litter, applied to excess manure across the basin. No adjustment is made for methane digesters, as byproduct nutrients are typically returned to cropland.

[10] Hauling costs were calculated based on an estimated per-ton and per-mile charge by manure type (lagoon, slurry, dry) and distance interval. Hauling costs are net of revenues received for purchased manure.

[11] An updated NRCS-CEAP assessment based on 2011 farmer survey data suggests rapid expansion in land-applied manure in the Chesapeake Bay watershed, with nearly half (48 percent) of farmland acres in the basin receiving manure.

[12] Here and in all subsequent analyses, scenario WTAM50 (assuming 50 percent of landowners are willing to accept manure) serves as a benchmark scenario for examination of alternative compliance scenarios.

[13] Conservation practices such as filter strips and riparian buffers may further reduce available acreage for manure spreading. However, the effect on manure hauling costs is likely to be very small on a watershed scale.

[14] Based on the nutrient content of poultry litter, averaging roughly 44 lbs. N and 22 lbs. P per dry-ton, with a third of applied P valued under an N standard. The value rises to $41/dry-ton when manure P is fully valued.

[15] Animal-units, as defined for use in the NRCS manure-nutrient database, represent 1,000 pounds of live animal weight. An animal-unit serves as a common unit for aggregating over different types of livestock.

[16] Net returns per animal-unit are based on the USDA-ERS REAP agricultural sector model.

[17] NRCS, 2003, Costs Associated With Development and Implementation of Comprehensive Nutrient Management Plans—Part 1.

INDEX

D

E

J

L

M

N

O

P

Q

R

S

T

U

V

W

Y